2nd IWA Leading-Edge Conference on Sustainability

WATER AND ENVIRONMENTAL MANAGEMENT SERIES

2nd IWA Leading-Edge Conference on Sustainability

Edited by M. Bruce Beck and Andrew Speers

IWA Publishing

Published by IWA Publishing, Alliance House, 12 Caxton Street, London SW1H 0QS, UK
Telephone: +44 (0) 20 7654 5500; Fax: +44 (0) 20 7654 5555; Email: publications@iwap.co.uk
Web: www.iwapublishing.com

First published 2006

Editing and layout by The Clyvedon Press Ltd, Cardiff, UK (www.clyvedonpress.com)

British Library Cataloguing in Publication Data
A CIP catalogue record for this book is available from the British Library

Library of Congress Cataloging- in-Publication Data
A catalog record for this book is available from the Library of Congress

ISBN: 978-1-84339-507-2
ISSN: 1476-1785

Contents

Preface

Sustainability is a paradoxical concept. We know we want to protect the environment from human-induced change, yet ecosystems are dynamic, constantly changing and adapting in response to a multitude of factors, the combined effect and subtleties of which are probably well beyond human calculation. Furthermore, our conscious desire to protect the environment – which forces us to think of humans as sitting outside ecosystems – conflicts with the unavoidable fact that we are an unconscious actor within those ecosystems.

We must also recognise that the goal of 'protecting the environment' is not a clear-cut objective. Perhaps, because of its complexity and propensity to change, we cannot know what the fully protected environment would look like. Individual preferences too make the conceptualisation of an ideal state impossible; do we strive for an ecosystem in which we play a minor part – barely influencing natural outcomes – or one that is more actively managed and provides for our needs or wants?

IWA is a member-driven association representing the water sector. Water is such a fundamental part our human/industrial environment and the ecosystems that surround us, that it is incumbent upon the Association to play a role in interpreting the concept of sustainability and promoting 'triple-bottom line' outcomes. Accordingly, as part of a broad programme, in 2001, IWA held its first Leading-Edge Conference on Sustainability. This book represents the outcome of the second conference held in 2004.

Neither this volume, nor the conference from which it draws, resolve the paradoxes described above. The papers presented here do, however, provide insight into the innovative thinking and practical projects undertaken across the globe that move us from patently unsustainable conditions to those in which our economic activity, impact on ecosystems and desire for positive social outcomes are in better balance.

IWA will continue its efforts to promote sustainability and interpret the concepts. This book is but a step on the way, but one that will hopefully be seen as a valuable contribution to the challenges confronting us as an industry.

Bruce Beck and Andrew Speers
December 2005

Sustainability — seen in Sydney

M. Bruce Beck

Wheatley-Georgia Research Alliance Professor and Eminent Scholar, Warnell School of Forest Resources, University of Georgia, Athens, Georgia, USA (E-mail: mbbeck@uga.edu).

From Venice to Sydney

LES Venice (2002) left me with two lasting impressions: a collective beating of breasts over our inability to define 'sustainability in the water sector'; but a much surer sense of what sustainability was not — that many solutions promoted as sustainable did not look so to that discerning audience. LES-04 in Sydney has left neither of these impressions. What was to be defined in Sydney, if anything, was an expansion of the triple bottom line: that water infrastructure solutions should not merely be environmentally benign, economically feasible, and socially acceptable, but also embrace the additional benefits of sustaining institutional and organizational learning. Some, Jon O'Riordan (British Columbia, Canada), argued a case in favor of this quadruple, or triple, bottom line being vaulted into a triple/quadruple *top* line. By this he intended that a 'society-centric' form of governance should evolve to replace the current 'state-centric' model, intervening to steer Society towards publicly and democratically defined priorities and goals — as framed by the triple bottom line. His society-centric vision would require a greater collaboration among the four institutions of modern governance: the public, corporate, civic-society, and knowledge (e.g., university) sectors.

So there has been significant progress since LES-02: much more specific detail in terms of practical, real-world experiences, including at the small scale of the individual household and the local community; the first empirical evidence of what constitutes a sustainable water utility; and feedback from utilities on indicators and metrics of sustainability. Vikki Uhlmann (University of Queensland, Australia) has developed a systematic approach to achieving less unsustainability in the performance of water utilities, where progress can be gauged against seven principles — ranging from balance over the triple bottom line, to least-cost planning, to adaptive management. Amongst utilities across OECD countries the desire is palpable: to move towards such indicators and consistent reporting of progress against them. The obstacles to realizing this desire are manifest in three of the other indicators: achieving Total Water Cycle Management (TWCM) in a watershed, which is rarely available institutionally to a utility; being participatory, for want of in-house skills in the social sciences; and real-value pricing of water, not least because of the challenge of incorporating the inter-generational equity inherent in the Brundtland definition of sustainability.

Unfortunately for what I have said of LES-04 in relation to LES-02, Vikki found that for her set of indicators to be deemed legitimate, utilities, regulators, and the public would first need to agree on a definition of sustainability! My personal view on this, born of my exploration of many definitions in the background document prepared for LES-02 (available from the IWA website: www.iwahq.org.uk), is that for all its vagueness Brundtland has yet to be bettered. Indeed, I cannot help wondering whether for this reason precisely it will prove to be as successful as Aldo Leopold's similarly vague land ethic of the 1930s/40s. And in what I have seen since LES-02, let me add, considerations of longer-term, inter-generational equity are most consistently conspicuous by their absence from approaches labeled 'sustainable'.

A very human dimension

The supreme achievement of LES-04 was the coherence of its 'story line', from start to finish; from considerations of the human dimension, through systems planning, leading-edge technologies, health, hygiene, and risk, new approaches to decision making (beyond cost-benefit analysis), and governance, but not, regretfully, to ecosystem functions and services. And in that omission, as we shall see, there is an important message.

Extraordinary attention was given to the intimate human dimensions of people's toilet habits — Saburo Matsui's (Kyoto University, Japan) discussion of their comfort levels when fulfilling their natural biological functions in the dry sanitation systems of Nepal and Vietnam, and the report of Björn Vinnerås (Swedish University of Agricultural Science, Uppsala) on the precise chemical and biological composition of our very personal outputs. Perhaps this was not the kind of human dimension any squeamish, professional social scientist might have had in mind. LES Sydney was a Conference primarily of 'literate engineers', let us say; engineers capable of stretching out of the confines of their own discipline/profession and reaching far afield. We were all reading the script from the same page. There was inter-disciplinary cross-over; the capacity for any disciplinary perspective to pop up in any presentation and its discussion. But we had to wait almost until the end of the Conference for Paul Jeffrey (Cranfield University, UK) to tell us: surveys show 'people do not want to participate'! What an arresting — and puzzling — thought indeed, in these days of stakeholders galore and the immense concerns to cater assiduously for their perspectives.

Imperatives for change

Economists put it this way: given the expected global population by the end of the Century, we are but half way to 'build out'. And it is not as if the world has spare freshwater and other resources lying idly about, waiting to be mobilized in the service of accommodating this expectation. Cities, in countries such as China, India, Brazil, and South Africa, will not develop over the next two decades in the way that they have over the past two decades, nor as cities might continue to grow in countries of the North. Their futures, and their future water-nutrient metabolisms and infrastructures, as drivers of the entire water cycle in watersheds, will not evolve as smooth extrapolations of past trends.

Future metropolitan infrastructures might need less of the brittle 'engineering resilience' and deeply embedded technological 'lock-in' we see today, and more of the 'ecological resilience' that endows them with 'biomimetic' capacities for damage limitation and self-repair, like the body of an organism. Cynthia Mitchell (Sydney, Australia) took this kind of resilience to mean visibility. Whereas Bazalgette and the other great municipal engineers of the nineteenth Century saw success in putting the offensiveness of sewage and sewers out of sight and out of mind, in tomorrow's world, she argued, metropolitan consumers will require blunt market signals to change their ways — houses and office blocks designed to 'let it all hang out', with their water and sewerage services on the outside for all to see. Those who invest personally in recovering all the P, N, and K resources from their personal outputs — instead of putting the wasting strain on the centralized wastewater treatment facility — deserve a financial 'resource bonus' in Thomas Dockhorn's (Technical University of Braunschweig, Germany) view. A household urine-separating device, with recovery and recycling of the thereby locally separated nutrients, ought to reduce that household's charges for remote, centralized wastewater treatment from €506/annum to €432/annum.

But then, what people really want in their daily lives are the services of clean dishes and illumination, not necessarily a supply of water or electricity thereto. We need neither ownership of the dish-cleaning device — it can be rented to us — nor necessarily water as the basis of the cleaning function. In short, there were hints in Sydney (in Simon Fane's (Sydney, Australia) presentation) of strategic change, from economies founded not on {producer–owner} relationships but on transactions of a {provider–lessee} type.

Introducing agriculture and climate change

Agricultural issues and the consequences of climate change, hardly evident in LES-Venice, and hitherto not mainstream concerns of the IWA, passed unremarked, as entirely natural constituents in the discussion at Sydney. We heard from Carol Howe (CSIRO, Australia) on the embodied water content of foodstuffs, which echoed the opening allusions (in my own presentation) to increasing pressures on water in the agricultural sector and the consequent global trading in virtual water. Holger Hoff (Potsdam Institute for Climate Impacts Research, Germany) reported on what is — to my knowledge — the first instance of a Global Climate Model (GCM) being used as a 'driver' of watershed behavior, that of the Jordan River (Israel/Jordan). Kai Tiedemann (University of Aachen, Germany) was searching for the means to re-afforest the arid coast of Chile, to harvest water from fog — but justified on the higher economic and political grounds of improving carbon sequestration for mitigation of climate change. The seedling trees require an artificial watering system until their heights attain a critical level, sufficient to penetrate the fog-bearing, atmospheric boundary layers with the appropriate wind speeds for successful self-irrigation from harvesting. One especially efficient means of catapulting thus the new forest onto a self-sustaining trajectory simply reaches back to the clay pots of the Incas of former centuries for its present-day technology.

Simplicity and indigenous knowledge

Oh, to be 'rural/agrarian' once again — is to be simple and sustainable, is it not? We have the one (simplicity), but not the other (sustainability), in countries of the South, do we not? From the hard-won experience of the North, we have some insights on being unsustainable. Have we not solutions then to being less unsustainable in the South; perhaps they should be 'membranes for all!', or some other such panacea? Now is the time for 'doing' not 'thinking'.

To all of this rhetoric, I say, nothing was simple, or intellectually undemanding, or methodologically trivial, or unsustainable, in Claudia Pabón Pereira's (Wageningen University, The Netherlands) study of nutrient-resource recovery and recycling in a small, remote Andean village in Peru. She simulated three alternatives, finding that the option including household urine-separation, community-wide anaerobic digestion of cattle manure, and a duck-weed pond for cleaning a cattle water supply, scored most highly on all three accounts of her version of the triple bottom line. It is 'unrealistic to talk about technologies that will perform well under any circumstances', she concluded. Others would say: there is little point in optimizing the part while pessimizing the whole. 'The inherent dynamic nature of a technological process' — she continued — 'and the fact that it is the product of the social context where it is developed and used' makes technology essentially 'place-based', as yet others would say.

As for me, I ask (rhetorically): what of the consequences a generation or two hence of providing immediate solutions today to the morally urgent problems of water supply and sanitation in so many parts of the world? Do we have time to think on this? Is asking the very question a moral obligation? Will the water infrastructures of the South be condemned 25–50 years hence to the same degree of technological lock-in we now bemoan (and denigrate) in the North?

The abstract, conceptual, even 'theoretical', insights distilled from understanding the complexity of the slowly evolved resilience in the coupled natural-human system in the Andean village — that village's Indigenous Knowledge (IK) — may tell us something of how to build greater resilience into large, metropolitan, people-water systems.

Cities, panaceas, and Heels of Achilles

The wheel never quite turns full circle, back to somewhere in the nineteenth century; before — we might wish — the water closet had become as we know it today; so that it could be re-shaped as a urine-separating device, and then put universally in place. Such a device might now seem the panacea (of source separation), like membranes, re-use, or decentralization (in their own, different ways). Each, as with the heel of Achilles, has its nagging little difficulties, there for no other purpose than to challenge us ever positively to further innovation. Just as water and diet for the individual on the upside of the household have a bearing on health and nutritional well-being, so they, together with that individual, personal health, govern success on the downside of resource recovery from our biological residuals. Separation of urine is great for nutrient recycling, not so for its burden of pharmaceutical residues. Water flows downhill, from supply as potable on the upside to environmental return on the downside; John Langford (University of Melbourne, Australia) raised his hand to note, alas, that wastewater re-use within the city implies the consumption of additional energy to push water somewhere back uphill. Decentralization of metropolitan wastewater treatment foregoes the upheaval of sewers and is consistent with visibly placing environmental responsibility into ever more localized pairs of hands; but who is to 'police' the decreasingly apparent actions of so many more of the responsible (irresponsible?) hands, or maintain the increasingly complex functions of the household?

It is one thing to endure a failure in your personal entertainment center in your household, quite another to put up with a failing, in-house, sewage nutrient-recovery device. Being 'rural and small' in the South may be richly complex; being 'urban and small' in the North may be riskily complex. There are promising strategic paths forwards and upwards — to the sunlit uplands of less unsustainability — but no once-and-for-all solution. 'Always learning, never getting it right'.

Water — calling the shots

As 'water people' we suffer at times not only from introspection, but also subservience. Cities evolve. Land is occupied, devoted to a given use, covered, uncovered, then covered again, and so on. And all the time what happens to the city's water environment will be buffeted about by the dictates of how land is used. Water has had to 'dance to the tune of land uses'. We, the IWA constituency and our colleagues elsewhere, do what we can, to make the best of the poor hand water has been dealt. Andrew Speers (IWA, Australia) and Paul Jeffrey would have it otherwise. They imagine re-arranging land uses — including in the rural surrounds of urban centers — to 'dance to the tunes of water resource availability and water quality status in time and space'. Cities, like the watersheds in which they sit, will be players in a cascade of water uses, with urban form progressively being adapted to this end, one land-parcel at a time.

Fundamental science: all just too up-close and personal

Yet the real issue in achieving sustainability in the metropolitan water sector is fundamentally about arresting the propagation of pathogens. When so much of the discussion focuses on nutrient — even common salt (as in our diet) — cycles, this can be lost to view. For as long as people must live and work near to people in cities, preventing the propagation of disease remains essential. And in that respect it is not so obvious that we currently have all the technologies for moving towards the less unsustainable systems we would seek for recovering clean water and enriched nutrients. Pathogens prosper in biofilms, presumably, therefore, on the surface of the separating barrier stopping their physical movement. Their defense mechanisms therein appear to be built in some way upon cell signaling: 'pathogen raising unto pathogen the alarm'. Our responses could be two-fold: create no biofilms; and disrupt the signaling system to disarm pathogen defenses. In this, we should look to furanones, urged David Garman (Environmental Biotechnology, Australia). But shall we then be obligated to look elsewhere, to find a solution to the (now unimagined) downsides of these substances, their production, and the agents we shall fire at the thus disabled pathogens?

'Waste equals food' we hear, with an appeal to the marvels of Nature's achievements in its natural biological and ecological endowments from 3.8 billion years of evolution — and juxtaposed with our equivocation on biotechnology. Humanity, we expect, will continue to press ever more in upon humanity — and pigs, poultry, ponds, and ducks. Around the tightly-knit nutrient cycles of the indigenous village 'ecologies' of communities in South-East Asia propagate pathogens too, with a propensity for zoonosis; the capacity for a (mutating) viral infection to shift from causing illness merely in birds, for example, to illnesses in mammals and human beings. And this, we know (from Saburo Matsui), is the rub: another Achilles heel, with here an unequivocal benefit to be had from future progress in the fundamental science of genomics, as much as from the elementary separation of urine and feces at source, vacuum trucks for the latter, the recovery of fertilizer from the former, and the fermentation and disinfection of solids returnable to the land.

When the numbers should not add up

There is a tendency for the wetting front of engineering enumeration to move outwards to embrace ever more of the triple bottom line (*mea culpa* too, in some respects!); a yearning to find the scheme of quantitative weights that somehow will allow us to add up all the incommensurate, qualitative, value-laden, subjective elements in the matrix of indicators of 'sustainability', to generate the single number — telling us to choose this design over that for a more sustainable activated sludge unit. Some things simply cannot be poured into the number-crunching device. And we know it; but not the line that separates systematic engineering analysis from democratic debate amongst stakeholders. Nor should that line necessarily remain fixed. There is something in mathematics called fuzzy programming. What strategy maximizes the *possibility* of satisfying the bundle of all our constraints and goals, in all their incommensurate imprecision (of the natural language in which we have expressed them)? This was the question to which Linda Cobiac (University of Queensland, Australia), the lone mathematician at LES-04 (as far as I know), was addressing herself.

Bottom line

One of my lasting impressions of LES-04 (Sydney), notwithstanding our being just literate engineers presuming to know something of other people's professional back yards (especially those of ecologists and social scientists), will be this: the ease of traveling there across disciplinary boundaries, in a manner whereby the rounded whole could be readily assembled from the parts. This is rarely — very rarely — achievable. And when it is, the numbers, for us, unfortunately do not add up. Surveys show: too few people want to participate in inter-disciplinary IWA activities.

On my flight back to the US I watched a news feature about how to solve our wastewater problems. It spoke only of the Living Machines® of Ocean Arks International. Neither LES-02 nor LES-04 witnessed any mention of these. Is it then that what the ordinary man or woman in the street (or on the plane) knows of this, is decidedly *not* what we in the IWA are talking about?

Freshwater supply and self-irrigation of plants in water-limited environments by fog precipitation

K. Tiedemann

Research Institute for Water and Waste Management, Mies-van-der-Rohe-Strasse 17, 52056 Aachen, Germany

Abstract The technical implementation of fog harvesting was successfully introduced in the 1980s. Until today, though, further research on optimization of catchment systems and amplified use of the technique has not been focused. This paper presents some outcomes of a pilot field study in Ecuador and on the Canary Islands and describes the core ideas of a follow-up reforestation project that is currently in preparation.

Keywords Rural water supply; irrigation; fog harvesting

Introduction

Fog harvesting as a source for drinking water was first described by missionary Bartolomé de las Casas in his reports on Indian history in the 17th century (de las Casas 1986). The technical implementation of fog water supply was rediscovered in the 1980s, following the pioneer investigations of Schemenauer and Cereceda (Schemenauer and Cereceda 1988) and has since been established in various countries, above all along the western coast of South America (Figure 1).

Figure 1 Location of fog harvesting hot spots: 1 – Hawaii, USA; 2 – San Francisco, USA; 3 - Ecuador/Peru; 4 – Chile; 5 – Canary Islands, Spain; 6 – Namibia; 7 – Cape Town, South Africa; 8 – Oman; 9 – China; 10 – Tasmania, Australia (altered after United Nations Environment Programme 1998)

Reports on the implementation of a catchment system in Chile amount to a water yield of typically 1–10 l per day per square meter catchment net (Schemenauer and Cereceda 1994). Another fog harvesting example from Swakopmund, Namibia, provided a freshwater supply of up to 2.7 l per day and square meter catchment net during the favourable months of August through January (Mtuleni *et al.* 1998). Enough drinking water to solely support the needs of adjacent small communities if multiplied by the total net surface of 10–20 catchment systems of approximately 50 m^2 each. The success of such pioneer projects led to a rising degree of popularity – with the result that the Environment Programme of the United Nations Organization (UNEP) published a guideline for the construction of a catchment system as shown in Figure 2 and Figure 3 in 1998 (United Nations Environment Programme 1998). Various parameters to support a high water yield of such catchment systems were defined, but so far the recommendations for an assessment of harvesting potential still give room for improvement. A general problem in this complex is the lack of comparability of fog events due to unsatisfying measurement techniques.

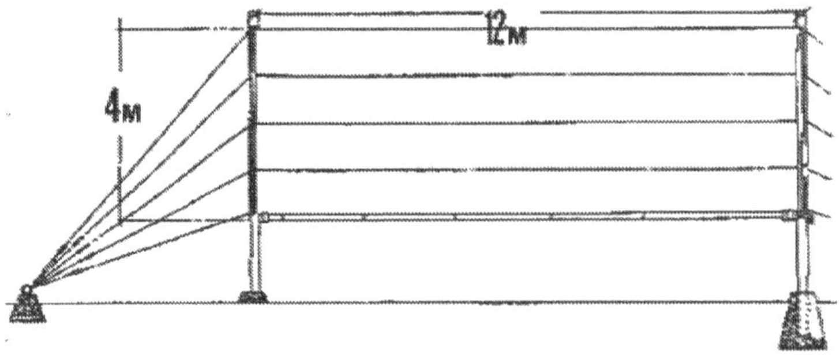

Figure 2 Construction sketch of catchment system (United Nations Environment Programme 1998)

Figure 3 Net structure type Raschel 50

Ellenberg (1959) published his observations on a Eucalyptus grove in the Peruvian Atacama Desert, which apparently covered most of its water request by its own fog precipitation (→ 'self irrigation'). Again, this report has never been validated scientifically, although it seems reasonable to assume, that this phenomenon might offer an irrigation option for fog deserts, which can also be found in Africa (Namib Desert), southern California or southwestern Australia.

Having this background in mind, the questions remain 'whether or not' and 'how' can the use of fog drip be technically optimized and be put into other fields of application. The research presented in this paper was designed to lay the scientific fundamentals for further pilot studies.

Materials and methods

Setting

The field research was conducted at two different locations. Investigations on the effect of plant structure on fog drip were carried out on Tenerife, Canary Islands, situated approximately 400 km off the northwestern African coast at 29° N. Tenerife's particular climate is dominated by the effects of persistent wind patterns, the surrounding cold ocean current, and the island's impressive altitude of more than 3.700 metres above sea level (m.a.s.l.).

Measurements on the self-irrigation effect of plants took place at the second site of investigation, the Bilsa Biological Station. Bilsa lies in the Mache-Chindul mountain range of Western Ecuador at an altitude of 350–750 m.a.s.l., some 50 km inland from the Pacific Coast just below the equator. Meteorologically, the area shows characteristics of lowland rainforests as well as of tropical montane cloud forests with a pronounced dry season, giving the unique opportunity to compare the influence of forests on effective precipitation depending on fog or rain conditions at the same site.

Depending on the season, fog occurs very frequently at both sites. On Tenerife, meteorological data show a 37.5% maximum percentage of hours of 100% relative humidity during spring months (Marzol *et al.* 1996). Similar long-term investigations in Bilsa are lacking due to poor meteorological equipment, but personal measurements have shown a percentage exceeding that of Tenerife during the dry season of 1999/2000.

Experimental Design

Water yield potential of plants and tin models

The term water yield potential was defined for these studies as the water yield by fog drip collected underneath test objects under optimal exposure. Nineteen plant species and three tin models with particular structural features have been investigated with respect to these patterns. The experimental design was simple: similar-sized branches of shrubs and trees, as well as tin models with surface areas of 1.600 cm^2 each were installed at a height of approximately 1.5 m above ground and exposed to wind-driven fog. A bucket was fixed underneath the objects to collect fog drip. The quantity of precipitation was measured in intervals according to fog conditions: favourable fog events permitted short measurement intervals of 10–30 minutes, while light fog accounted for overnight observations.

Self irrigation on permanent plots

Two 10 m × 10 m plots were cleaned of secondary growth and reforested with sixteen saplings of *Croton lechleri* (average height approx. 30 cm) and *Cordia alliodora* (average height approx. 40 cm) each. Within each plot, two precipitation collectors were installed, plus another collector outside the reforested area as a reference. From November 1999 until October 2000, precipitation data of the five collectors was collected daily at 6:00 p.m., plant height was measured once a month.

Results

Water yield potential

The precipitation measurements showed remarkable tendencies that led to the finding of five determining factors for the objects' water yield potential:

1. compact plant architecture: densely constructed branches may shift air flow and inhibit the exposure of potentially fog-collecting surfaces;
2. filling of space: a regular distribution of fog-collecting surfaces enhances fog drip efficiency of branches that allow for a good air flow;
3. leaf structure: leaf roughness may increase the leaf's surface tension and thus enhance droplet run-off;
4. leaf stability: leaves that are attached loosely to the branch can easily turn sideways with the wind to minimize their flow resistance, thus exposing only a fraction of their surface;
5. leaf edge-surface ratio: small leaves show a better fog drip efficiency than large ones due to a higher leaf edge-surface ratio, since leaf edges appeared to be the main site of droplet coagulation.

Deviated feasible incitements for structural improvements of catchment systems are a three-dimensional catchment cube and a catchment harp instead of the currently used catchment nets.

Self-irrigation

The outcomes as shown in Figures 4 and 5 are clearly a result of two meteorological phenomena: rainfall interception (= evaporation loss directly from the tree canopy) and fog drip. Both quantities increase as the plants grow higher and the canopies become wider. On the one hand this is due to the increase of Leaf Area Index (LAI = square meter leaf surface per square meter ground covered by the canopy); on the other this is due to a threshold height of objects above ground that is needed to reach fog-laden air masses with sufficient wind to support fog drip. According to meteorologists, wind speed increases significantly with height above ground, being hardly measurable at heights lower than 50 cm (Häckel 1990). This corresponds well with the observation of approximately 80 cm being the minimum tree height to show a systematic difference between precipitations during days of rain (precipitation ≥ 2.0 mm/d) or fog (precipitation < 2.0 mm/d). Generally, a considerable self-irrigation effect can thus be expected to set in at plant heights of 1 m and above.

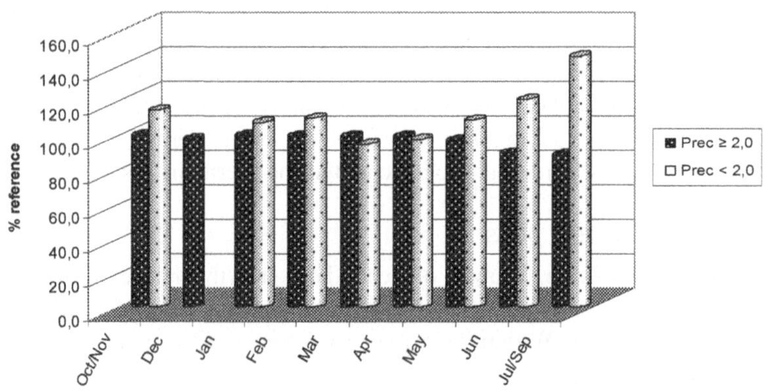

Figure 4 Percentage of precipitation within reforestation plots in relation to the reference

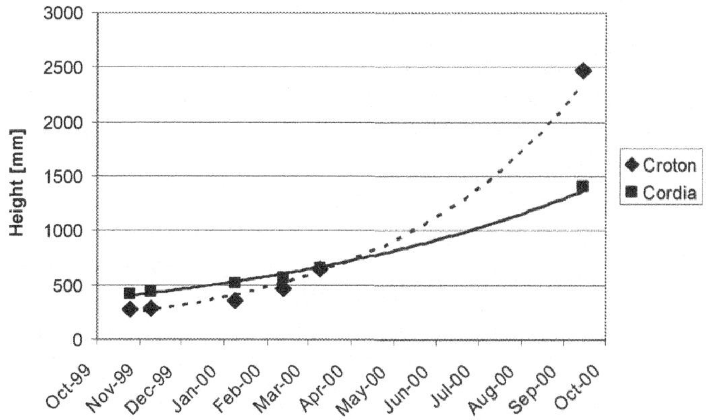

Figure 5 Plant height of *Croton lechleri* and *Cordia alliodora*

Impact and importance

Project 'Green Desert'

The results presented above offer a scientific platform for a future afforestation project in fog deserts that is currently in preparation. The main focus of this project is to amplify the use of fog harvesting systems from sole drinking water supply to a simple and comparatively cheap alternative to surface water irrigation in water-limited environments.

The project aims at meeting triple bottom line outcomes (see brackets) by working on the following compounds:

- installation of a three-dimensional catchment cube and a catchment harp as innovative systems for irrigation purposes and enhanced freshwater supply of adjacent communities (→ society);
- suggestions for an irrigation system during the initial afforestation phase (→ environment);
- implementation of an integrated farming model (→ society & economy);
- an evaluation of appropriate plants with regard to structural features for enhanced water yield by fog precipitation as identified in the pilot study to activate a sustainable hydrological cycle (→ environment);
- an evaluation on the feasibility of desert reforestation projects for carbon sequestration purposes and emissions trading (→ economy).

Structural improvement of catchment systems

As mentioned earlier, catchment systems have been successfully installed in fog-abundant regions all over the globe. However, fog harvesting applications have not yet focused on possible structural improvements of the catchment nets. As the results in the pilot studies have shown, the water yield has proven to be particularly high under objects with a pronounced three-dimensional structure as well as under objects with comparatively few horizontal disruptions. Consequently, two approaches to an enhanced water yield can be offered:

- to dwell on the use of the same net type as shown in Figure 3, but connect 10–15 nets to a three-dimensional structure or
- to implement a new type of catchment objects by introducing an array of vertical strings and thus avoiding any horizontal disruptions.

Both approaches will be tested with regard to cost–benefit relations.

Integrated farming in water-limited regions by implementing small-scale irrigation systems

The western coast of sub-equatorial South America is characterized by extreme aridity in connection with a high abundance of fog events. These particular meteorological patterns account for the opportunity to develop an integrated farming model in regions that hardly provide any other source of income for its inhabitants.

In the early stages of the reforestation project, catchment systems as shown in Figure 2 are needed to irrigate well adapted tree species to a height of roughly 1 m, where the self irrigation effect is expected to set in. Research on simple irrigation systems have been carried out for example by Ngigi, Thorne, Waweru and Blank (Ngigi *et al.* 2001). Figure 6 shows the suggestion for a cheap installation that distributes water from a 200 l tank to an afforestation plot of 150 m². Another option is the implementation of pottery irrigation as shown in Figure 7, an ancient technique that originates from the Inka times. By using sub-surface clay pots as an individual water storage for each tree or crop, water is made available as the plants need it, hence water savings between 50 % and 70 % (compared to the conventional watering system) can be achieved. Additionally, no salt accumulation occurs in the soil and the spreading of weeds is suppressed as the surface soil layer remains dry throughout the growing season (Daka 2001).

Once the grove has reached a state of independence from external water sources, the catchment systems can be used either for further plantations or for the sole use as a source of drinking water in small scale communities. The self-irrigating plot will be used for undergrowth crops that benefit from the pioneer trees' fog precipitation. Again, these plants must be well adapted to shady and semi-dry conditions: a selection task for local experts in agroforestry from the Lima University, which is one of the partners in the development of this project.

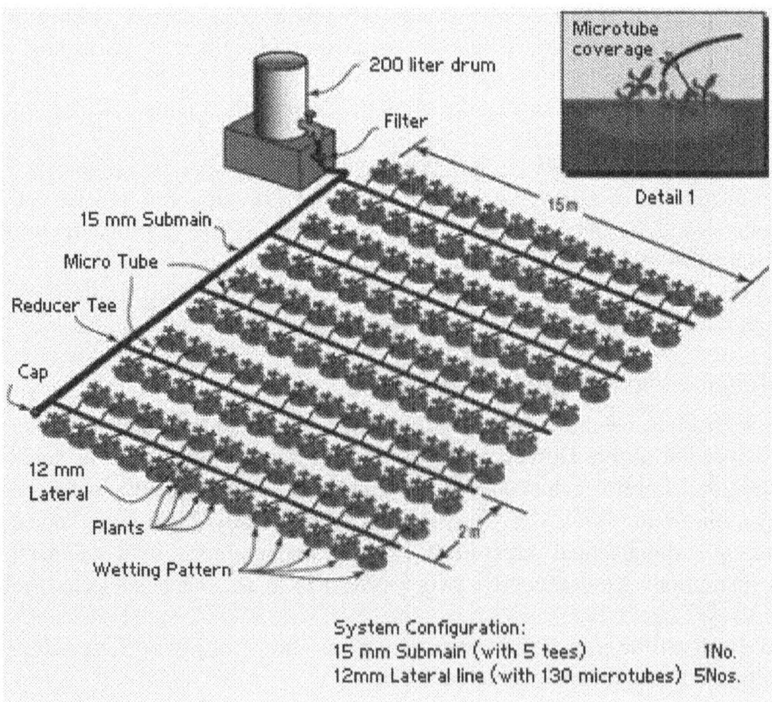

Figure 6 Example for a simple irrigation system (International Development Enterprises 2003)

Figure 7 Clay pot operation principles (Daka 2001)

Carbon sequestration and emissions trading

After the Kyoto Protocol to the United Nations Framework Convention on Climatic Change was adopted in 1997 by the participating parties, national and international laws and measures had to be developed to assure the successful implementation of the United Nations' goal to reduce atmospheric carbon dioxide concentration. European governments recently passed the Joint Implementation and the Clean Development Act, and have thus paved the way for developing countries to evolve a new perspective to natural resources: an economic value of enduring forests as a biomass store. Furthermore, afforestation measures can be labeled as biomass fixation with the opportunity to sell emission equivalents to interested parties from industrialized nations.

As simple as the idea seems, a sound calculation of the amount of sequestrated carbon dioxide may be challenging. Pioneer projects in South America realized the carbon sequestration concept by enrichment of existing disturbed secondary forests. However, particularly in complex ecosystems such as tropical lowland rainforests, extensive research has to be carried out to clearly separate the effect of afforestation from the natural dynamics of the forest, since claim holders can only tender a confirmed surplus biomass that entirely attributes to their engagement.

Apparently, these efforts are considerably smaller in less complex environments. From this perspective, fog deserts offer an ideal basis for afforestation measures, as initial biomass inventories are fairly manageable due to a comparatively low degree of biodiversity and a small number of plants. Hence, it seems reasonable to assume that with regard to cost–benefit relations a comparison of the starting positions *lowland rainforest* on the one hand and *fog desert* on the other hand should generally favour the latter.

Conclusions

Regardless of the technique's simplicity, fog harvesting has shown to be a complex matter in the presented context. Considering the challenges of the 21st century in water management and water supply, applications beyond the current state of fog water use and further research are strongly indicated. The *green desert project* covers a variety of disciplines related to fog harvesting (see Figure 8). Although final outcomes are still uncertain, yet the project offers an approach to integrated water management of potentially great importance.

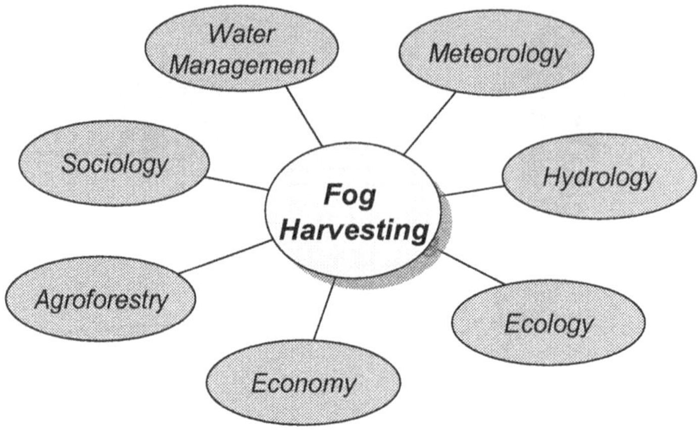

Figure 8 Disciplines related to fog harvesting

References

Daka, A.E. (2001) Development of a technological package for sustainable use of dambos by small-scale farmers. PhD thesis, Dept Plant Prod. Soil Sc., Univ. of Pretoria, South Africa.

de las Casas, B. (1986) Historia de las Indias. 2nd Edition, Biblioteca Ayacucho, Caracas, Venezuela.

Ellenberg, H. (1959) Über den Wasserhaushalt tropischer Nebeloasen in der Küstenregion Perus. *Berichte über das geobotanische Forschungsinstitut Rübel 1958*, 47–74.

Häckel, H. (1990) Meteorologie. 2nd Edition, UTB, Stuttgart, Germany.

International Development Enterprises (2003) Drip Irrigation. Internet document downloaded at http://www.ideorg.org/html/gallery/drip.html (24.04.2004).

Marzol, V. *et al.* (1996) La captación del agua del mar de nubes en Tenerife. Método e instrumental. *Clima y agua: la gestión de un recurso climático*, La Laguna, Spain, 333–350.

Mtuleni, V., Henschel, J.R. and Seely, M.K. (1998) Evaluation of Fog-Harvesting Potential in Namibia. In: Schemenauer, R.S. and Bridgman, H. (Eds.) Proceedings of the First International Conderence on Fog and Fog Collection. IDRC, Ottawa, Canada, 179–182.

Ngigi, S.N., Thorne, J.N., Waweru, D.W. and Blank, H.G. (2001) Low-cost irrigation for poverty reduction. International Water Management Institute, Annual Report 2000-2001, 23–29.

Schemenauer, R.S. and Bridgman, H. (Eds. 1998) Proceedings of the First International Conference on Fog and Fog Collection. International Development Research Centre, Ottawa, Canada.

Schemenauer, R.S. and Cereceda, P. (1994) Fog Collection's Role in Water Planning for Developing Countries. Meteorological Service of Canada, internet document downloaded at http://www.msc-smc.ec.gc.ca/armp/fog/old/unnrf.html (13.05.2000).

United Nations Environment Programme (1998) Sourcebook of Alternative Technologies for Freshwater Augmentation in Latin America and the Caribbean. *Technical Publication Series* **8c**, Osaka, Japan.

Socially fair and environmentally sound developments in the programme for rural water supply on the Lesser Sunda Islands (Nusa Tenggara Timur), Indonesia

W. Kirchhof* and P. Minnigh**

*Research Institute for Water and Waste Management at the Technical University of Aachen, 52056 Aachen, Germany
**c/o ProAir Project Office, FC team leader, Waingapu, Nusa Tenggara Timur, Indonesia

Abstract This paper summarizes the results from the first 2 years of the 8-year Indonesian-German cooperation programme rural water supply on the Lesser Sunda Islands (NTT). The programme emphasizes on the rehabilitation and construction of water supply facilities for about 100,000 inhabitants. The methodology based on an open demand-driven approach with community participation. The project areas covered mattock regions with low groundwater sources. The social rural system is characterized by population layers with an upper class of population and groups of serfs. The share of women employed in the modern public and private sector is low. Two German consultant groups assist the Indonesian Health Government to improve the health conditions in NTT delivering technical services, implementing environmentally sound water exploration and treatment methods and management concepts for water supply systems with economically fair standards.

Keywords Rural water supply; mattock regions, Indonesia; demand-driven approach

Introduction

Geographical and social data

The province of Nusa Tenggara Timur (NTT), the lesser Sunda Islands, covers approximately 80,769 km². The islands stretch 2850 km from west to east (115°49′ E to 134°54′ E) and to south 11° S (Figure 1). The small island of Pamana is the most southerly island in Indonesia. Kupang, the provincial capital, is 800 km from Darwin, north Australia.

Figure 1 Map of the Lesser Sunda Islands

The Lesser Sunda Islands form one of Indonesia's most economically disadvantaged provinces. This is indicated by large deficiencies in the qualitative and quantitative water supply of the population. The lack of supply with uncontaminated water and a low level of awareness for hygienic issues are the main causes for major health risks. The rainfall patterns are characterized by a long annual dry period; the islands are widely dominated by mattock regions (Figure 2) with low groundwater resources.

The present population is estimated at 3.5 million with an annual growth rate of 1.3% in the 1990s. Rural exodus was estimated at 0.31% for that decade, but it is said to be grossly underreported. At present the average population density is about 74 inhabitants per km², varying from district to district between 23 inhabitants per km² (Eastern Sumba) and 150 inhabitants per km² in Sikka (Eastern Flores).

Ethically the province is extremely diversified. Dozens of different languages and several hundred dialects are spoken. This is the result of complex prehistoric and historic immigration patterns, settling and ruling structures as well as of centuries of fighting between many small and some bigger local chiefdoms and kingdoms. More than half of the inhabitants are Catholics, one third are Protestants, only about 8% are Muslims. Since 1985 a so called translocation programme has been launched. Villagers living in remote and highland areas were asked to move to the lowlands since the infrastructure is much better there.

Figure 2 Aerial view of Sumba Island, east of Waingapu, in dry season 2004 (photo taken by author)

The economy on Sumba is widely based on barter transaction. Therefore, sincere problems with the money economy occur. The exchange of money between members of a family or within a family group is difficult. People create reciprocal relationships by lending money or effects, and by not directly giving them back. The fact that they are in debt means that their relationship with the other is strong and that both have to take care of each other in order for the debt ever to be paid back. Exchanges with effects, textiles and or food, are more common.

Figure 3 Photo of typical village on Sumba Island (photo taken by author)

Historically, the communities perceived water supply infrastructure as a gift from the government. Under these conditions any efforts towards community empowerment to take over operation and maintenance (O&M) after installation have been neglected. According to a survey in 2000 the main technical and community based deficiencies seemed to be:

- Inadequate design and pipeline routing, lack of valves
- Low–quality cheap materials and fittings
- Poor construction techniques and quality
- Pipelines laid too close to the surface and thus easily damaged
- Pipelines frequently exposed to landslides and flooding
- Lack of involvement of recipient communities' ineffective management in rural districts. The responsibility was passed to other persons or organisations and after a while nobody felt responsible. Ownership has never been really transferred.
- Lack of corresponding training for user groups or proper networking between neighbouring villages and willingness to assist each other.

New national water policy

The process to formulate a Community-Based Water Supply and Environmental Sanitation National Policy began in 1998. The Government of Australia through AusAID has provided technical assistance in the Water Supply and Sanitation Policy Formulation and Action Planning (WASPOLA) program, to WSP-EAP (Water and Sanitation Program for East Asia and the Pacific).

This policy was prepared in stages by a Water Supply and Environmental Sanitation (WSES) Working Group led by the National Development Planning Agency/ BAPPENAS. This initiative incorporated participatory approaches; it entailed wide-ranging collaboration and consultation with key stakeholders in the WSES sector such as local governments, donor agencies, non-government organizations, academia, and others. In line with the spirit of the decentralization and regional autonomy policies, the WSES Working Group has accommodated inputs from local governments in the effort to determine how a national policy can be implemented at the regional level. Several local governments were selected and have participated in a trial to comprehend the process of implementing a national policy.

Other equally significant contributions for the policy finalization were gathered from several case studies of the WSES sector and from trials involving the field application of the policy principles in large projects funded by loans from the IBRD and grants from KfW and UNICEF.

In essence, this policy emphasizes the need for a shift in the paradigm for WSES development into one that focuses on the sustainability and effective use of WSES infrastructure and services. The policy also outlines the basic strategies necessary to implement the policy principles such as how to help apply the demand-response approach, improve human resource capacity, raise community awareness, promote environmental sanitation, develop institutional arrangements, and strengthen the monitoring and evaluation schemes during all phases of the WSES development process. The next crucial step is to formulate a national action plan that can be applied in both the long and medium term, as well as on an annual basis.

Project set-up

Since 2002 the German Government provides financial (FC) and technical assistance (TC) in the Rural Water Supply Development Programme on the Lesser Sunda Islands (NTT). The Indonesian-German co-operation programme aims to rehabilitate and construct water supply facilities for about 100,000 inhabitants on the Lesser Sunda Islands (NTT).

The project methodology based on the implementation of an open demand-driven approach. The programme is set to give full responsibility to community groups to develop themselves with the support of projects and the government, but not driven by those. The groups have to be responsible, and ready to take over the management, operation and maintenance of the systems. They have to organize themselves and have to pay for water in whatever way they can.

The organizational structure of the programme (Figure 4), decided at the Indonesian Ministry of Health, shows the strong linkages between the organizations of central and district levels. The complex network of national and provincial organizations gives only small margins to develop the demand-driven approach that is founded on decision-making by the village communities.

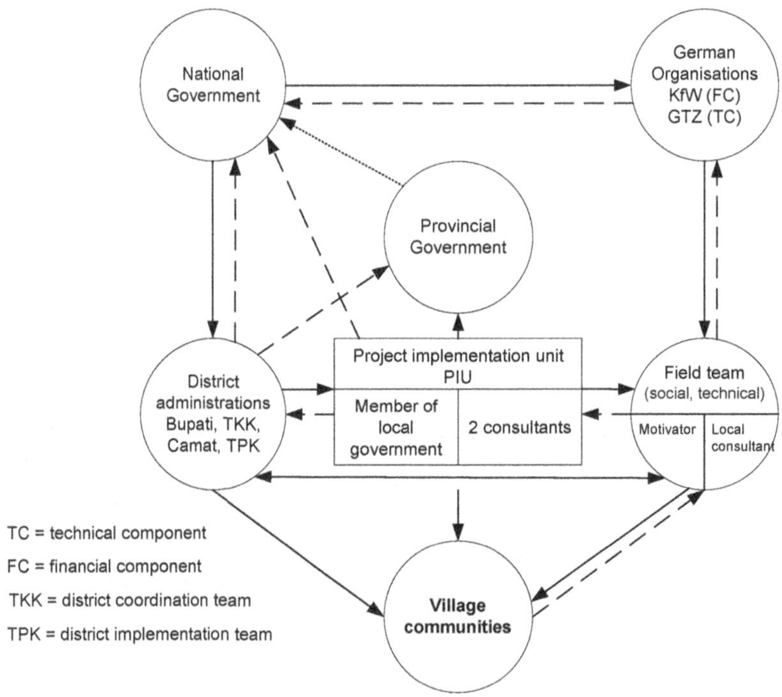

Figure 4 Organizational structure of the programme with linkages to the acting parties (modified, after DGCDCEH, 2004)

The TC component makes a socially fair decision for the construction of water supply schemes, establishment and empowerment of water user associations, strengthening of district administrations to support and supervise local water supply schemes, introduction of tools for the protection of water resources, and the planning and construction of pilot sanitary schemes.

The FC services comprise the development of an implementation concept, preparation of information campaigns, assistance in application procedures and selection of locations, tendering of local consultants for detailed planning and local construction works, supervision of construction and commissioning.

The impacts model of the programme (Figure 5) indicates the strong inter-linkages between the desired achievements and benefits.

Project execution

Since July 2002 the consultant developed the long-term programme implementation plan, prepared information campaigns, implemented a dissemination mechanism, support the communities in preparing applications, evaluated the programme locations in compliance with the technical and social feasibility, supported the preparation of technology concepts for informed choice, increased the acceptance of communities into the programme's principles, and selected local consultants, contractors through competitive bidding processes. Figure 6 shows the status of sub-tasks in selected locations at the mid of 2004.

Conclusions and lessons learned

A. Communications

It is difficult to find new group leader in these rural areas. People in general say 'saya tidak bisa suruh keluarga maupun teman', 'I cannot command or ask others relatives or friends to do something'. It means it is hard to be in charge as nobody will automatically accept leadership, except from people who are already the leaders. This is hard if you have to run a community organization, have to organize labour etc. It becomes even harder that if you have included the local leaders that most of them have other priorities and less time for the activities you have in mind for a water project. As a result most of the organizations include traditional leaders, but only a few can function in a flexible way. Things are changing one's mind which are normally a matter of the local regulations, are now harder, as people cannot change their local leaders. There are ways around it, by appointing people in an honourable position, and moving them out of the way, but to start up an activity these movers or pioneers are needed. Therefore to include them is difficult; to exclude them is not possible.

The village community members could be mobilized by initial information campaigns. Information campaigns were held in the sub-districts and the villages, only explaining about the existence of the project, the basic principles, the approaches, and the fact that the community groups themselves had to make the request directly at a certain time in the project office. The reaction of the community was overwhelming; of 22 villages in East Sumba we received 84 requests. It shows that there is a lot of initiative in a village and people are willing to go the extra mile to obtain what they think they would like to have.

The social team needed technical understanding of the water supply systems before conducting interviews. During the first intake interviews the technical team was not present, as it was expected that their role would begin later with technical surveys based on the requests. However, this proved to be not so, first of all, the social team needed technical understanding of the situation at the locations. They had problems in general about technical issues including perceiving and making small situation sketches, and understanding technical aspects of existing systems.

The social team members had to learn how to do constructed interviews. The local staffs were not used asking questions and really making an effort to see cause effect relations, and why a certain situation arises. Questioning other people is often deemed very impolite in Indonesia. As a result most of the information obtained via these first intake interviews was useless.

The communication was improved with posters explaining technical aspects. After oral explanations of the different alternatives the technical team started to use posters. A general poster about a gravity fed system explained technical aspects of such a system. One poster explained the possible service levels, and the fact that this will mean that people have to decide among themselves who will be served by one terminal. One poster depicts directly the situation at the location, in the village at the different zones. The posters of the zones show not only the pipe line, the sources and the planned structures, overlaying the area based on the existing topographical maps, but also the houses and the public buildings.

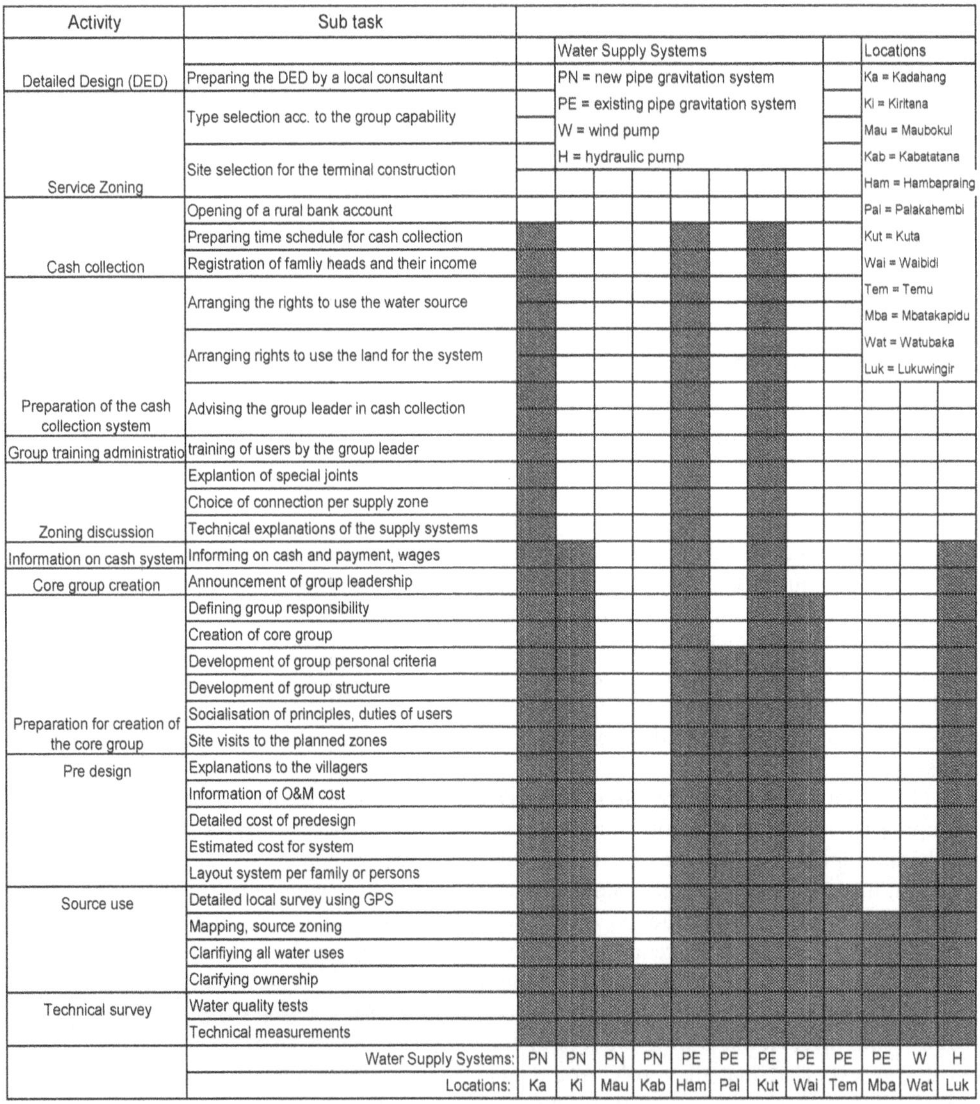

Figure 5 Impacts model of the programme with linked achievements and benefits

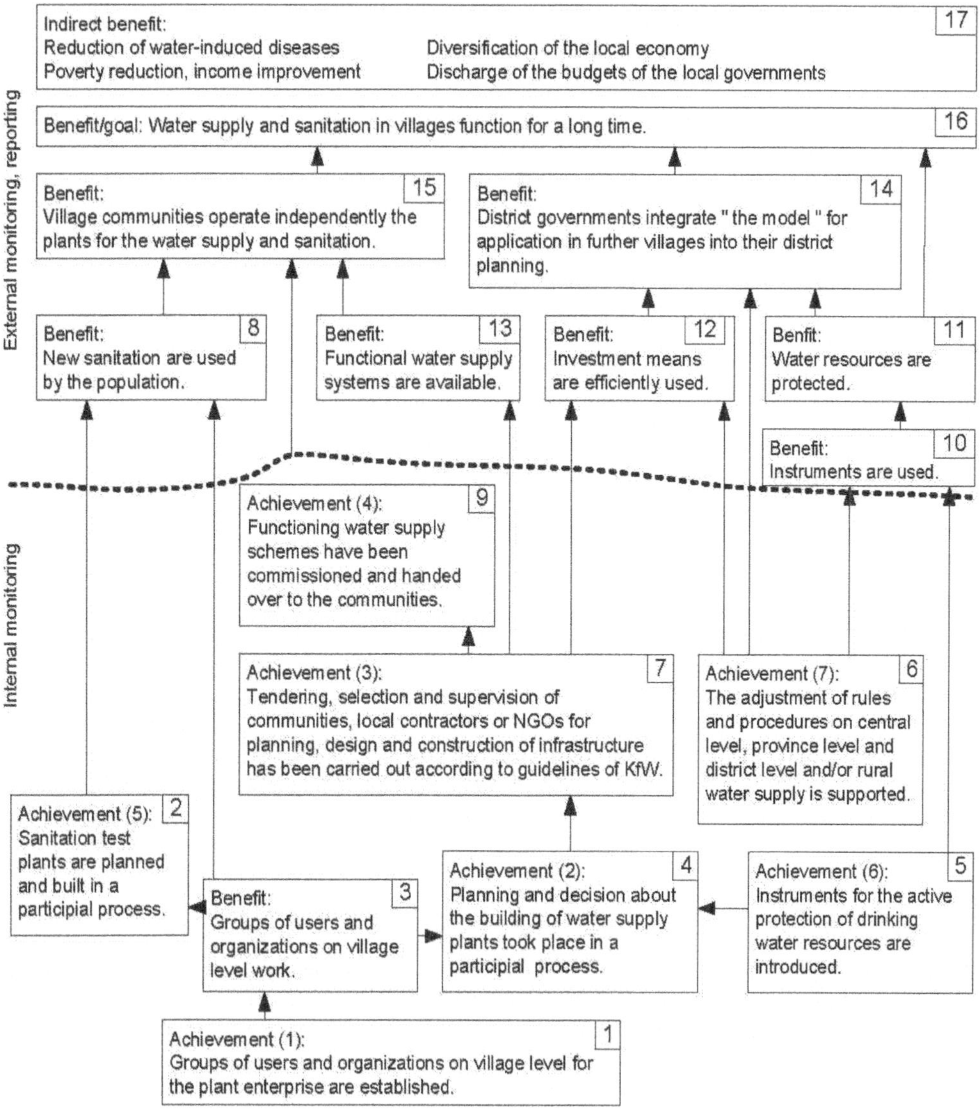

Figure 6 Activities, sub-tasks and their levels of realization at the project sites

B. Technology

The request criteria for site selection have to set limits. To avoid a large number of requests, the minimum size of community was changed from 10 families to 30 families. First priority was given to requests for a gravity fed system, with an existing source. However, we still got a number of requests for other systems. This mainly occurred because the group who proposed the system actually consisted of individual families or small groups of families asking for a number of small systems like dug wells, rainwater-catchment reservoirs and others. As we needed to focus on one system, we chose to first follow-up on the gravity fed systems, and left the smaller system to a new round.

All pre-designs of the water supply systems, which were made for three different sizes, have to be explained. Starting in Sumba Timur, the technical team engineers informed the communities about their findings and the possible alternatives the community groups would have in choosing a water system. With the gravity fed systems (GFS) they designed three alternatives, a low, medium and high priced system, whereby the difference was mainly made up by the difference in pipe diameters and the volume available to the community members. When they explained this in words and figures, all community groups listened and choose for the largest system. It was clear that their philosophy was very simple, they wanted the biggest pipes with as much water as possible, and did not regard the follow-up costs. Nobody provided sufficient arguments, and a real commitment was not reached, as no money was paid for the in-cash amount. It took the team several months to see that nothing happened and that changes were necessary.

The number of terminals in a water supply system must be limited. In large water projects it was seen as positive community participation if after the system had been build, the community members themselves installed and added other connections, often house connections. This is in technical terms the same as destroying of the system. To avoid this, it was decided that the design of a system include a maximum number of terminals. A terminal is defined as a water outlet, which can be transformed in a public tap, house connection or reservoir, or other. A terminal is meant to serve a maximum of 60 people for a minimum of 40 litres per person per day.

The service areas were divided into zones with different service levels to ease the control. Based on the topography, and hydraulic gradient, zones were defined using as much as possible the administrative boundaries of hamlets. All zones have a maximum allowance of terminals. Every zone gets one branch, and the total number of terminals is connected to that point. If people make more tap-points, this should be considered as criminal as they will damage the system and the water supply of other people. However, the terminals can have different service levels, in terms of distance of the houses to the tap, and volume per day, so in choosing wisely people will be able to match their needs with the system design.

C. Institutions

'The community should be the driving force of the project' is the most challenging project goal. The people themselves have never been given this chance, and are used to wait and ask for help. They have cultivated this way, and until now nobody sees why they should be responsible or active themselves. Naturally, there are groups and NGOs who are active and take destiny in their own hands but they are still few. This initiative is often brought to a halt because the government takes over in support of the poor, thereby neglecting and even ending all local initiative.

The technical, social team members and motivators had no experience with a demand driven approach for a community based water project. In the beginning the social team and technical team had to be engaged in their own preparation and learning process. For the social team this included a selection training of the motivators. The selected people from the area had university degrees but no experience as there are hardly any opportunities for work except for the governmental positions. The team leaders faced the fact that they had to build up the team from scratch. Several local specialized NGOs assisted in this task and the project recruited several people who already had experience in other projects. However nobody had experience with a demand driven approach for a community based water project, as devised under the project contract, and in line with the ideas expressed in the WASPOLA policy. Until now, it had been merely theory for the situation in Indonesia, except for a few mainly small-scale NGO projects.

It was advisable to split the community into smaller groups, often called focus groups. The teams saw that in talking with the community they always talked with different people and it was not clear who were in the group who had a real interest. Therefore it was necessary to work in smaller groups, which are often called focus groups. However, though focus group discussions were part of the MPA, it was very difficult for the young inexperienced motivators to hold these meetings because they did not know how to split up the groups.

Concerted efforts, called INFOKUS, of all teams were necessary to improve the villagers' acceptance of the principles. In a joint action, the technical and social teams together with resource persons went to villages for about 3 days. In three days, two are spend by groups in the zones, and one is used for a panel discussion. The joint team explained about the system designed for them based on their request, tried to get people interested and knowledgeable about this system via the posters, tried to get people together to form an organization, or if that organization is already formed, to get it more widely accepted and its function more widely known.

D. Financing

The barter economy hampers the cash flow. The people should pay for water, to barter is not sufficient. The project has now the task to find out if the people themselves can find a way in which to convert goods into money. Otherwise it might be helpful to have others help with this transition.

The project is not serving the very poor. We have realized that working on a water supply project; we do not in first instance support the poor as such. All water systems demand an investment and demand that there is operation and maintenance to be able to keep the system in a good state. Therefore, only families willing and able to contribute can be considered. The overall effect will be that most of the families in an area will benefit, as local customs and social behaviour will in general accommodate evident differences in wealth.

The project does not accept subsidies. The project waits until the community is ready to pay the in-cash contribution of 4% of the investment budget for materials, based on the pre-design figures. This is hardly acceptable to the Sumba Timur government, as they have paid already in full in several of the first villages who make up the first round.

The technical team advised the choice of the most optimal system for the community, which is still affordable. Affordable is defined as that the in-cash contribution should not be above Rp. 25.000 (EUR 2.50) per person (total investment for materials divided by total number of beneficiaries), and that the fee to be paid every month for O&M should not be over Rp. 2000 (EUR 0.20) per person per month. With salaries in cash of between 10 and 100 EUR per family per month, this was considered to be still affordable. Optimal has been defined in the largest pipe dimension possible so as to make available the largest volume of water possible, in order for the people to get as much water per person per day as close as possible. The joint team informed the people about the necessary in-cash amount to be made available before construction can start.

Conclusions

The implementation of an open demand-driven approach in rural water supply programmes is a new approach in Indonesia. Therefore, the project that should give full responsibility to community groups to develop themselves is hampered by obstacles in the field of communications, technology, institutional building and financing matters.

It is difficult to find group leaders in eastern Indonesian rural areas because of existing governmental regulations and tradition rules. An effective mobilization of village community members demands initial information campaigns with technical and social topics. The motivators and facilitators have to be well trained to get a sufficient high level technical understanding of the water supply systems before conducting field surveys and interviews. It is very advisable to divide the whole village area into smaller service areas with different service levels to ease the control. It is advisable to split the community into smaller groups, so called focus groups.

As the barter economy hampers the cash flow and the project does not accept subsidies, it is advisable to assist village communities in their choice of system so that they can chose the most optimal system for the community, which is still affordable.

References

Directorate General Communicable Disease Control and Environmental Health (2004) Pedoman Umum, Program Air Bersih dan Sanitasi Pedesaan (Guidance document, Rural Water Supply and Sanitation), Ministry of Health, executing agency, Jakarta.

Central Project Management Unit (CPMU) (2004) Water and Sanitation for low income communities project, General Project Description, Ministry of Health, Directorate General Communicable Disease Control and Environmental Health, Jakarta.

Water Supply and Sanitation Policy Formulation and Action Planning (WASPOLA) Project (2003) National Policy, Development of Community-Based Water Supply and Environmental Sanitation, National Development Planning Agency/BAPPENAS, Jakarta.

Assessing water service provision scenarios using the concept of sustainability

S. Maheepala[*], M. Evans, A. Sharma[***], S. Gray**** and C. Howe[*****]**

*PO Box 56, Commonwealth Scientific Industrial Research Organisation, Highett, Victoria 3190, Australia. Shiroma.Maheepala@csiro.au
**Montgomery Watson Harza, Hamilton, New Zealand. martyn.t.evans@nz.mwhglobal.com
***PO Box 56, Commonwealth Scientific Industrial Research Organisation, Highett, Victoria 3190, Australia. Ashok.Sharma@csiro.au
****PO Box 56, Commonwealth Scientific Industrial Research Organisation, Highett, Victoria 3190, Australia. Stephen.Gray@csiro.au
*****PO Box 310, Commonwealth Scientific Industrial Research Organisation, North Ryde, New South Wales 1670, Australia. Carol.Howe@csiro.au

Abstract There is an increased interest in seeking alternatives to the traditional way of providing urban water services, particularly in areas where new developments are planned. Whilst the exact reasons for such an interest vary from case to case, in general, the underlying reason is to provide a service that is more sustainable than the traditional practice. One of the major challenges faced by the water industry when examining alternative water servicing options is lack of an assessment method that can embrace the principles of sustainability. This paper addresses this challenge and presents a method for the assessment of alternative urban water servicing options based on the concept of sustainability. The proposed assessment method uses multi-criteria assessment method to integrate social, economic and environmental aspects of decision-making. It is applicable at the planning and conceptual design stages of the water system of both greenfield and infill development schemes. Briefly, it requires a set of sustainability assessment criteria and at least one measurable indicator for each criterion be defined, values of indicators to be estimated for alternative scenarios and ranking of scenarios using a multi-criteria assessment method. Applicability of the proposed method has been demonstrated using 750 hectares of Greenfield development in Papamoa in New Zealand.

Keywords Urban water, sustainability, sustainability assessment, multi criteria assessment

Introduction

The traditional approach to managing urban water has been to exploit surface and ground water sources, create dams to supply towns and cities and then dispose of the wastewater and stormwater back into inland waterways or the ocean. The main objective of this approach is to fully meet *all* water demand and protect public health. The price that we are paying today for enjoying the meeting of all water demands is an erosion of the ecological integrity of both source and sink environments such as aquifer depletion, surface water exploitation and degradation of receiving water bodies. Sustainability of the traditional approach is now being questioned, and alternatives to the traditional approach are being actively sought (Victorian Government Department of Sustainability and Environment, 2004; Pinkham, 1999). At present, there is increased encouragement from many water authorities and governments to examine the viability of integrating alternative sources (e.g. stormwater and recycled wastewater) into the existing water supplies.

Often many alternative service provision options offer multiple benefits. For example, benefits associated with stormwater use for non-potable uses include direct benefits such as reduced demand for potable water supply and delay or removal of the stormwater drainage infrastructure, and indirect benefits such as protection/enhancement of receiving waters through reduced stormwater flows and contaminant loads (Hatt *et al.* 2004), which has the tendency for an increase in the recreation value associated with receiving waters and enhancements in property values of nearby properties. One of the major challenges faced by the water industry with regard to examining alternative water servicing options is lack of an assessment method that can capture these multiple economic, social and environmental benefits and costs associated with water servicing options.

Researchers in the Commonwealth Scientific Industrial Research Organisation, Australia (CSIRO) have addressed this challenge by developing a method for sustainability assessment, which is particularly applicable at planning and conceptual design stages of alternative urban water systems in both greenfield and re-development areas. This paper provides a brief description of the sustainability assessment framework and an application of it to a greenfield development in New Zealand.

Sustainability Assessment Framework

Sustainability assessment is a process by which the implications of an initiative (e.g. proposed or existing policy, strategy, project, etc.) are evaluated to determine the extent to which the initiative contributes to *sustainability state,* defined by a goal and a set of criteria derived from the principles of sustainability (Pope *et al.* 2004). Sustainability assessment allows practitioners to identify major steps towards achieving a defined state of sustainability. Often these major steps represent significant changes to the current state, thus requiring a considerable capital investment.

The sustainability assessment methods reported in the literature includes the method developed as part of UK's SWARD project (Foxon *et al.* 2002), CSIRO's method (Maheepala *et al.* 2003), the method developed as part of Swedish MISTRA project (Hellstrom *et al.* 1999). All these methods follow a similar structure. The main features include setting objectives and a set of assessment criteria based on the principles of sustainability, defining of a set of measurable indicators under each assessment criterion, use of multi-criteria assessment method for ranking of alternatives and carrying out of this whole assessment process with a group of stakeholders. In this paper, we describe the sustainability assessment framework of Maheepala *et al.* (2003), which has been specifically devised for assessing the sustainability of urban water servicing options.

The sustainability assessment framework of Maheepala *et al.* (2003) can be used to: (1) compare alternatives to the traditional method of providing urban water services and (2) measure the sustainability of a given urban water system, which can be either the current urban water system or any other preferred urban water system. The Assessment Framework requires ten steps to follow and these ten steps are:

Step 1: Define sustainability assessment criteria
Step 2: Define measurable indicators for each criterion
Step 3: Identify the base case and/or generate options
Step 4: Collect data and estimate values for indicators
Step 5: Decide on a multi-criteria assessment method to be used
Step 6: Assess options using multi-criteria assessment
Step 7: Conduct sensitivity analysis
Step 8: Communicate the assessment process and outcomes to stakeholders

Step 9: Receive feedback from stakeholders
Step 10: Finalise assessment and choose a preferred scenario for detailed design
The processes involved in each step are briefly described below.

Stepl involves defining the 'desired state' using a set of criteria.

Step2 involves defining measurable indicators under each criteria category. It is required to define at least one indicator for each criterion. There is no maximum limit to the number of indicators that can be used under a criterion. By considering practical and operational use, it is recommended that only a few indicators should be used for each criterion.

Step 3 involves identifying the base system for which sustainability assessment is to be conducted. The base system can be the current system without any changes or any preferred system against which alternative systems are compared. Identification of the base system also requires defining of boundaries as well as components of the system. Once the base case is identified, a broad range of alternative options can be generated if the objective of the assessment is about selecting the most sustainable option. The options generated would reasonably be expected to meet the objectives, but might be substantially different from the base case. On the other hand, if the objective of the assessment is to determine the sustainability of the base case, there is no need to generate options.

Step 4 involves collection of data to estimate values of chosen indicators corresponding to the base case and each option (if there are options), and estimation of values of indicators using the available data and interpretive methods/models. Data should also be collected to define target values for each indicator. The target values are the best values that could be expected if the system under consideration is sustainable. It should be noted that currently, there is very limited information available on target values and certainly, further research is essential to understand target values associated with the sustainable urban water systems.

Step 5 involves deciding on a MCA method. The recommended MCA methods are PROMETHEE (http://www.visualdecision.com), and ELECTRE (Lamsade Software, Lamsade, Universite′ Paris-Dauphine). Both these MCA methods are available as commercial software, and require subjective information such as weights of importance of each criterion and preference value for each criterion. Determination of the subjective information usually relies on the use of expert panels or selected water industry stakeholder groups. To obtain subjective information, a workshop type of approach is recommended. After discussing the issues at the workshop, some method of voting should be adopted that allows input from each participant to be included in an overall evaluation of the weighting of each issue (or indicator) and an assessment of their relative weights against each other.

Step 6 involves an assessment of sustainability using the chosen MCA method. This is a straightforward calculation since both PROMETHEE and ELECTRE software are designed to perform outranking of option(s) based on the both subjective information and quantitative information of the chosen indicators.

Step 7 involves performing of sensitivity analyses to obtain a robust solution. Often this process involves varying the values of subjective information by trial and error and assessing the sensitivity of subjective information on the outranking and the final solution.

Step 8 involves having discussions on inputs and outputs of the analysis with the stakeholders, preferably using a workshop type approach.

Step 9 has been devised to incorporate feedback from stakeholders into the decision process. Depending on the level of feedback, the decision process should be repeated from Steps 1 through to 9 by refining the criteria, options, indicators, data collection, methods used for estimating values of indicators and subjective information. This iterative process must be continued until all or a majority of stakeholders are satisfied with all the inputs and outputs of the decision process. Once all (or a majority) of the stakeholders are satisfied, the decision process proceeds to Step 10.

Step 10 involves finalisation of the assessment outcomes. At this point, a decision is being made as to which should be the preferred option if the assessment is about determining the most sustainable option. Alternatively, a comparison can be made to a selected option (or the base case) with target values of indicators if the assessment is about determining the sustainability of a selected option (or the base case). In the former case, the preferred option is the option that outranks all other options (i.e. the option at the top of the ranking list).

Case Study – Papamoa development in New Zealand

The Papamoa development in New Zealand was a residential and commercial development of 750 hectares of rural land in Papamoa East, in the jurisdiction of the Tauranga City Council (TCC). The proposed development had to meet the following site-specific objectives:

- Lower the impact on the natural environment
- Must be socially acceptable
- Must be economically viable
- Minimise health risk to the community
- Must recognise Maori cultural values
- Must be specifically applicable to the study area

The above-mentioned sustainability assessment framework was applied to Papamoa East development to determine a water servicing option that best meets the site-specific objectives. An application of each step of the framework is described below.

Step 1: Define sustainability assessment criteria

Guidelines provided by Hellstrom *et al* (2000) and the stakeholder input were used to define a set of assessment criteria in line the site-specific objectives. The assessment criteria adopted for this study were:

1. Economic criteria – defines the desired state in terms of economic costs and benefits associated with urban water service provision
2. Social criteria - defines the desired state in terms of social needs and expectations which include health requirements and maintaining inter and intra generational social equity
3. Environmental criteria – defines the desired state in terms of enhancing ecosystem wellbeing and protecting hydrologic integrity of the urban water system
4. Health criteria – defines the desired state in terms of community health aspects.
5. Cultural criteria – defines the desired state in terms of cultural values, i.e. Customary Maori values in this study.
6. Technical criteria – defines the desired state in terms of technical feasibility, which takes into account technical and/or engineering factors that influence applicability of the option and any governance issues.

Step 2: Define measurable indicators for each criterion

Indicators considered under each criterion and the objective or the purpose of each indicator are summarised in Table 1.

Table 1 Sub-assessment criteria used for the case study

Main Criterion	Sub-criteria or indicators (ID)	Objective of sub-criteria
Economic	Total infrastructure cost (EC1)	To maximise the economic viability of water service provision
Economic	Cost to customers (EC2)	To minimise the cost to individual customers
Social	Social acceptance (S1)	To maximise the social acceptability of water service provision
Environmental	Freshwater use (E1)	To minimise resource depletion
Environmental	Energy use (E2)	To minimise the use of fossil resources and related greenhouse gas emissions
Environmental	N discharge to receiving waters (E3)	To minimise contamination of water ways
Environmental	P discharge to receiving waters (E4)	To minimise contamination of water ways
Environmental	TSS discharge to receiving waters (E5)	To minimise contamination of water ways
Environmental	BOD discharge to receiving waters (E6)	To minimise contamination of water ways
Environmental	Wastewater generation (E7)	To minimise wastewater generation
Environmental	Stormwater generation (E8)	To mimic pre-development stormwater flows
Environmental	Recycling of nutrients (E9)	To maximise nutrient use locally thereby reducing the use of fossil resource and GHG emissions.
Environmental	Use of material for construction of infrastructure (E10)	To minimise material use
Environmental	N discharge to land (E11)	To minimise contamination of soil and land degradation.
Environmental	P discharge to land (E12)	To minimise contamination of soil and land degradation.
Environmental	TSS discharge to land (E13)	To minimise contamination of soil and land degradation.
Environmental	BOD discharge to land (E14)	To minimise contamination of soil and land degradation.
Health	Public Health water supply (H1)	To minimise health risk of water supply
Health	Public Health wastewater (H2)	To minimise health risk of wastewater provision
Cultural	Customary Maori values (C1)	To ensure customary Maori values are reflected appropriately, and do not damage heritage items in the course of provisioning water services and its operation.
Functional	Out-leakage (F1)	To minimise unaccounted water in water supply
Functional	In-leakage (F2)	To minimise wet-weather flow in sewers
Functional	Odour (F3)	To minimise air pollution
Functional	Flooding (F4)	To minimise frequency of flash flooding.
Functional	Landscape and regulation (F5)	To maximise the use of existing landscape features and the current regulatory frameworks.

Step 3: Identify the base case and/or generate options

Six alternative water service provision options (or scenarios) were considered and they are summarised in Table 2.

Table 2 Alternative water service provision scenarios considered for the case study

Alternative Water Service Provision Scenarios
The Status Quo (Basecase) – Scenario 1 This scenario comprises all the tools and techniques that reflect the current form of infrastructure provision and may form the 'Basecase' against which other scenarios may be assessed. For water supply and wastewater this is predominantly a traditional 'pipe in – pipe out' technique, however current measures TCC undertake to encourage reducing water consumption are included. The stormwater infrastructure is configured to reflect the current use of 'on-site' soakage pits for roofs where soils permit and the end point of disposal for paved areas and roads to the Wairakei drainage reserve.
Sustainability Improvement Model 1 - Scenario 2 This scenario takes all of the tools and techniques from the Basecase and welds on the following additional techniques: compulsory household water saving devices (dual flush toilets, low water use showers, washing machines etc), introduction of limited 'alternative' water sources (rainwater ranks for garden use and/or groundwater if the latter is available), supply of treated wastewater (i.e. reclaimed water) for irrigation of parks and reserves and industry use, provision of neighbourhood stormwater soakage pits when on-site soakage not available and greater use of swales for road drainage and other techniques to reduce run off and contaminants reaching waterways.
Sustainability Improvement Model 2 – Scenario 3 This scenario builds upon the tools and techniques used in Scenario 2 and introduce the following additional techniques: rainwater tanks are used at a household scale as peak levelling tanks (i.e. topped up from reticulated supply); a ban is imposed on selective uses of water (e.g. car washing on driveways); treated wastewater (i.e. reclaimed water) is made available for public non-potable use (e.g. garden irrigation); a tiered rate structure is used for wastewater disposal; management of truck movements imposed in commercial areas (e.g. dedicated hard standing, wash down areas) and stormwater is treated in wetlands prior to disposal to waterways and / or the ocean
The Mauri Model includes the two main Tools/Techniques that were proposed in the paper submitted by MMP to the SmartGrowth forum, namely: *'Te Pae Tuatahi: Human Effluent Without Water'* and *'Te Pae Tuarua: Catch and Use Rainwater'*. The Scenario comprises the following main components: on-site rainwater tanks for garden and laundry use; a reticulated potable water supply but at a reduced scale; composting (waterless) toilet and all the stormwater components of Scenario 3 and a neighbourhood car wash facility. This model was however sub-divided to address three differing methods for disposing of the greywater component of wastewater due to varying implications on the proposed analysis. • **Scenario 4a On site disposal of all greywater.** • **Scenario 4b Off site disposal of greywater via piped network to centralised wastewater treatment plant.** • **Scenario 4c 50% of greywater disposed of on site and 50% off site.**

Step 4: Collect data and estimate values for indicators

Monitoring strategy of each indicator is shown in **Table 3**. Values for some indicators were determined quantitatively through modelling of water and contaminant flow paths and costs. Qualitative assessment was carried out, for cases where sufficient data were not available. The estimated data values of the each indicator for all six scenarios are summarised in Table 4.

Table 3 Indicators and the monitoring strategy adopted

ID	Monitoring strategy adopted
EC1	Quantitative using life cycle cost analysis - Life cycle cost of infrastructure, over a defined planning period (50 years)
EC2	Qualitative – Based on the charges paid by customers for urban water services and costs involved in maintaining infrastructure provided at property level
S1	Qualitative – Based on an adequate and safe water supply being readily available throughout the year, wastewater disposed of in a safe and cost effective manner and properties not effected by flooding unless due to an extreme event
E1	Quantitative using water balance analysis - Annual amount of fresh water imported to the study area in ML
E2	Qualitative – Based on the annual electricity use by the total water system
E3	Quantitative using contaminant balance analysis - Annual N load discharge to receiving waters in tonnes
E4	Quantitative using contaminant balance analysis - Annual P load discharge to receiving waters in tonnes
E5	Quantitative using contaminant balance analysis - Annual TSS load discharge to receiving waters in tonnes
E6	Quantitative using contaminant balance analysis - Annual BOD load discharge to receiving waters in tonnes
E7	Quantitative using water balance analysis - Annual wastewater flows leaving study area in ML
E8	Quantitative using water balance analysis - Annual stormwater flows leaving study area in ML
E9	Quantitative using water balance analysis - Annual volume of treated wastewater recycled in ML
E10	Qualitative – Based on the total length of pipes and concrete channels used for urban water transportation
E11	Quantitative using contaminant balance analysis - Annual N load discharge to land in tonnes
E12	Quantitative using contaminant balance analysis - Annual P load discharge to land in tonnes
E13	Quantitative using contaminant balance analysis - Annual TSS load discharge to land in tonnes
E14	Quantitative using contaminant balance analysis - Annual BOD load discharge to land in tonnes
H1	Qualitative – Based on delivering water in compliance with A/NZ Drinking Water Quality Guidelines.
H2	Qualitative – Based on not placing public health at risk due to use of alternative treatment and disposal options plus re-use of reclaimed wastewater
C1	Qualitative – Based on the likelihood of preserving customary Maori values
F1	Quantitative - Annual volume of water likely to loss via leakage
F2	Qualitative – Based on the number of sewer over flow incidents per year at property level and distribution system level.
F3	Qualitative – Based on the number of odour complaints per year.
F4	Qualitative – Based on the likelihood of not increasing the frequency and intensity of flash flooding.
F5	Qualitative – Based on the likelihood of utilising the existing landscape features, soil characteristics and compliance with regulatory frameworks.

Table 4 Data values of indicators used for sustainability assessment

ID	Unit	Scenario					
		1	2	3	4a	4b	4c
EC1	$M	125	146	166	190	190	210
EC2	5-point	3	2	2	1	2	1
S1	5-point	3	3	2	0	1	0
E1	ML/yr	847	792	163	419	419	419
E2	5-point	2	2	3	3	3	3
E3	kg/yr	9544	9126	7069	4848	7203	5991
E4	kg/yr	1161	1109	821	655	886	766
E5	kg/yr	357795	329912	236445	274166	271698	270976
E6	kg/yr	70784	67503	48811	43381	52459	47641
E7	ML/yr	810	797	736	0	553	276
E8	ML/yr	2251	2183	1857	2169	2138	2139
E9	ML/yr	0	13	71	0	0	0
E10	5-point	2	2	2	3	2	2
E11	kg/yr	37	102	392	2960	37	1462
E12	kg/yr	1	7	35	828	1	413
E13	kg/yr	224	261	437	66739	224	33481
E14	kg/yr	112	362	1532	89636	112	44818
H1	5-point	3	3	2	3	3	3
H2	5-point	3	3	2	0	1	0
C1	5-point	2	3	4	4	3	3
F1	ML/yr	68	63	13	54	54	54
F2	5-point	2	2	2	3	2	2
F3	5-point	2	2	2	1	1	1
F4	5-point	3	2	3	2	3	2
F5	5-point	3	3	3	0	1	0

Step 5 and 6: Decide on a MCA method and perform assessment

The PROMETHEE MCA method was used as the MCA method, which was available as a commercial software system (Decision Lab 2000). The PROMETHEE MCA method required subjective information such as weights of importance of each criterion and preference function for each criterion. As mentioned earlier, determining of the subjective information usually relied on the use of expert panels or stakeholders of the study. A number of workshop was held to create scenarios, but a specific workshop to decide a weighting scheme for the final sustainability assessment criteria was not held, hence an 'equal weighting' system had been adopted among the six main criteria. That is, each main criterion was given equal weighting, i.e. 1/6 (or 16.67%). The weighting value given to each main criterion was then split equally among the sub-criteria (or indicators) of that main criterion. For example, the environmental criterion has 14 sub-criteria so the weight of the each environmental sub-criteria is 17%/14 = 1.2%, whereas the economic criterion has 2 sub-criteria so the weight of the each economic sub-criteria is 8.3%. In addition, in this study, we adopted a preference function without both indifference and preference thresholds, which implied that the decision maker considered any difference in values between two scenarios on any sub-criterion was decisive.

Table 5 shows ranking of scenarios from the best one to the worst one. The *phi* value associated with a particular scenario can be considered as the overall relative score of that scenario, and it is called *Phi score of a scenario* - the larger the *phi* value the better the scenario in terms of being more sustainable. Thus, the best scenario in terms of meeting the principles of sustainability is Scenario 3.

Table 5 Ranking of Scenarios

	3	2	1	4b	4a	4c
phi	0.27	0.20	0.17	-0.06	-0.15	-0.42

Step 7: Conduct sensitivity analysis

The sensitivity analysis of the final assessment ranking revealed that Maori cultural values, followed by economic and then social criteria, were the most sensitive to any change in weightings, i.e. changing their weighting would have the greatest impact on final rankings. Environmental criteria were the most robust.

Doubling the weightings of Maori cultural values still ranked Scenario 3 first, but when doubling the weightings of either social or health criteria, Scenario 3 ranked third. When doubling the weightings of economic criteria Scenario 3 is ranked second.

Steps 8, 9 and 10: Finalisation of sustainability assessment outcomes

The outcomes of the assessment process were presented to stakeholders and the feedback received did not change the weighting scheme adopted because there was no strong justification as to one assessment criterion was more important than another. Therefore, it was concluded that conceptually, preferred urban water service provision options for Papamoa East are those described under Scenario 3, closely followed by Scenario 2.

This concludes application of the sustainability assessment framework to Papamoa East. The next step is to carry out a detailed analysis for the preferred two scenarios and select a scenario based on that detailed assessment. Then an engineering design must be carried out for the preferred scenario before taking up any measures for implementation.

Conclusions

Sustainability assessment is emerging as a new tool for assessing and comparing alternative urban water servicing options. CSIRO Urban Water has developed a sustainability assessment framework by combining life cycle, multi-criteria and indicator assessment methods. The strong aspects of this framework are its ability to handle multiple conflicting objectives measured in various units and logical handling of subjective information. Both are essential aspects for measuring sustainability because of the value-based nature of the concept of sustainability.

This paper demonstrates applicability of the sustainability assessment framework developed by CSIRO Urban Water through a case study that examines the sustainability of six alternative water servicing options in Papamoa East in New Zealand. The assessment framework needs a set of criterion based on at least social, economic and environmental aspects of alternative water servicing options be considered, but other criteria can be considered in addition to the base criteria, if required, based on expectations and aspirations of the community. For the Papamoa East case study, three additional criteria were considered along with the base criteria and they were based on the public health, functional and the customary Maori cultural aspects. Since there was no strong justification as to one criterion is more important than another, all six criteria were considered as equally important, which led to the urban water servicing option described under Scenario 3 to be the conceptually, preferred urban water servicing scenario for Papamoa East.

Acknowledgements

This project was funded and data and information were provided by the Tauranga City Council, New Zealand. Authors would like to thank Andrew Speers (ex Director of CSIRO Urban Water Program) and Jim Bradley of Montgomery Watson Harza (New Zealand) for their valuable direction and Keni Piahana for his input with regard to customary Maori values.

References

Foxon, T.J., McIlkenny, G., Gilmour, D., Oltean-Dumbrava, C., Soulter, N., Ashley, R., Butler, D., Person, P., Jowitt, P. and Moir, J. (2002) Sustainability croteria for decision support in the UK water industry. *Journal of Environmental Planning and Management*, **45**(2), 285–301.

Hatt, B., Deletic, A. and Fletcher, T. (2004), Integrated stormwater treatment and reuse systems – an inventory of Australian practice. Monash University.

Hellstrom, D., Jeppsson, U. and Karrman, E. (2000). A framework for systems analysis of sustainable urban water management, *Journal of Environment Impact Assessment Review*, **20**, 311–321.

Maheepala S., Speers, A., Booker, N. and Mitchell, G. (2003) The Institution of Engineers, Australia, 28th International Hydrology and Water Resources Symposium, 10–14 November 2003, Wollongong, NSW.

Pinkham, R. (1999). 21st Century Water Systems: Scenarios, Visions and Drivers, An opening, presentation for an EPA Workshop on 'sustainable urban water infrastructure – a vision of future', Rocky Mountain Institute, Snowmass, Colorado, http://www.rmi.org.

Pope, J., Annandale, D. and Morrison-Saunders, A. (2004) Conceptualising sustainability assessment. *Journal of Environment Impact Assessment Review*, **24**, 595–616.

Victorian Government Department of Sustainability and Environment (2004) Securing our water future together.

The secret life of water systems: least cost planning beyond demand management

S Fane, A Turner, and C Mitchell

Institute for Sustainable Futures (www.isf.uts.edu.au), University of Technology, Sydney, PO Box 123, Broadway, New South Wales 2007, Australia

Abstract The water industry in Australia and international is involved in a period of significant change. The conventional roles of water and wastewater utilities are being redefined with the objectives of resource conservation and sustainable development added to existing responsibilities.

Least cost planning (LCP) has emerged as the way forward for water utilities in regions where water conservation has become an objective or where ongoing supply expansion is constrained. It involves techniques for the design and evaluation of demand management programs and aims to compare demand- and supply-side options on an equivalent basis. The approach is based on the key ideas that: demand is for the services water provides rather than the actual volume supplied; and that a drop of water saved is equal to a drop supplied.

This paper contends that LCP has much to offer the water sector beyond demand management. It is an approach that has potential for options assessment across the water cycle and can aid planning towards more sustainable outcomes within the sector. The paper concludes that LCP concepts and techniques will have worth in addressing the challenges of sustainable development for both urban water systems and catchment management.

Keywords Backcasting; economic analysis; end-use modelling

Introduction

Least cost planning originated in the energy sector in the United States during the 1980's for comparing energy conservation programs to increased generation as sources of supply (Beecher, 1996). It was based on the realization that a kilowatt of electricity saved through demand management was the equivalent of an increased kilowatt of supply. More recently LCP has emerged as a way forward for water utilities looking for ways to provide for new customers in regions where opportunities for the expansion of bulk supply are constrained. Least cost planning for water supply involves a suite of techniques for the design of demand management options and programs, the estimation of water conservation outcomes, and the evaluation of costs. These assessment techniques aim to identify the lowest cost means of providing customers with the water-related services rather than the water itself (Howe and White, 1999).

The conceptual distinction between supplying water services rather than water is critical to how LCP differs from conventional 'supply-side' approaches. Based on this idea, a water supply and distribution system should not be designed or managed in order to meet a simple projection of increasing bulk water demand, but should be designed to satisfy the water-related service needs or end-uses. The key focus of LCP is then on what is normally the 'secret life' of our water systems: how water systems interact with system users to deliver the water-related services that people actually want and how the demand for specific end-uses such as clean clothes, hot showers, amenable landscapes is likely to change in the future. As well as this focus on end-use services, LCP is characterized by the deliberate linkage of demand forecast to backcasting, and by a distinctive approach to option evaluation.

In the water industry the usefulness of LCP in the design and evaluation of demand management options has been recognised. Beyond demand management, however, LCP has much to offer the water sector as an approach for planning towards more sustainable outcomes. This paper describes the key LCP concepts and techniques and explores a number of ways these might be applied more generally and with further development LCP has a significant role as an approach for identifying sustainable outcomes for both urban water systems and catchment management.

Service provision and end-use modelling

Least cost planning is based on the understanding that demand for water is in fact demand for the services water provides. As an approach it shifts the focus of analysis from the quantity of water delivered to the quality of the service provided. This change means that the conventional rationale of continuing supply expansion is no longer appropriate. Instead taking an LCP approach water utilities should consider demand management, source substitution and bulk supply options in unison, with the aim of identify how best to provide water-related services to system users (the customers).

The focus of LCP is at the point of interaction between system users and water systems. This is where end-use occurs. Both the characteristics of the water using equipment employed and user behaviour are then of interest. The end-use of household sanitation, for example, can be provided by single flush 12 litre, dual-flush 9/4.5 litre or 6/3 litre and now 4.5/3 litre dual-flush toilets or even with no water by a composting system. The amount of water required to provide household sanitation also depends on how the equipment is used. For example, do users flush both before and after use and is the half flush facility utilised?

The foundation of assessment in LCP is end-use analysis and end-use modelling. These methods epitomize the principle that demand is for water-related services rather than bulk supply. End use analysis and modelling disaggregate water demand into individual sectors (i.e. residential, commercial, industrial, institutional and unaccounted for or non revenue water), and end uses within these sectors (i.e. toilets, showers, clothes washers). End-use analysis builds up a picture of water use through customer meters, surveys of water using equipment (i.e. showers, clothes washers, toilets) and water using practices (i.e. frequency of shower use and clothes washing), and evaluation of industry sales statistics from manufacturers (White, 1998). End-use models combine this technical and behavioral data with demographic and land use change projections to produce detailed forecasts.

An end-use model forecasts both the future demand for water services and the resulting bulk supply demand. Detailed end-use modelling provides a much more rigorous basis for demand forecasting than simple demand trend models and this improves estimates of timing and costs of conventional supply augmentation (White and Fane, 2001). End-use models also allow the water savings from demand management and other conservation measures to be estimated. Both water efficiency and source substitution options can be better designed and evaluated using end-use models. Increased water efficiency can be achieved by either replacing water using equipment in existing properties or regulating the sales of new equipment. Source substitution options such as rain tanks and effluent reuse can be targeted at specific end-uses. Examples include, roof water being used for the laundry and garden watering and greywater diversion for toilet flushing. The focus of LCP on end use services allows options for providing the same end-use with less water or lower grade of water to be explored.

Forecasting and backcasting

Backcasting studies involve describing a desired future end-point and then working back from that point, to determine the feasibility and what would be required to reach that goal. Backcasting is useful when problems are complex, there is need for major change and/or dominant trends are part of the problem (Dreborg, 1996). It is an approach for exploring means by which specified future states might be attained (Robinson, 1982).

Forecasting in comparison involves projecting into the future based on what are perceived as the current dominant tends. Forecasting conceives of the future as immutable and a derivative of the present and the past while backcasting addresses the potential for people to change significant aspects of the future as it occurs. The future is of course both of these things, being in part derived from the present and in part the result of deliberate shaping.

Mitchell and White (2003) contend that forecasting and backcasting are complementary in planning for the future of water systems. Forecasting can tells us important information about the near future. This allows us to optimise the existing system in the short term. Backcasting is in contrast much more powerful as a means of reflecting on the medium to long term. This is because it allows assumptions about how systems might be configured in the future to be challenged and reconsidered.

Detailed forecasting is the basis of insightful backcasting. Since forecasting in LCP is based on understanding the underlying trends driving the demand for water (demographic change and the need for water-related services) rather than crude predictions based on the past volumes supplied, it opens the potential for backcasting. It is this kind of careful identification of which variables represent the underlying tends and which variables are mutable to future action that characterises a powerful backcasting study. The focus on service provision in LCP allows the source, quality and quantity of water used for each end-use to be variables in the analysis.

Least cost planning studies can focus on either forecasting or backcasting. Studies can use LCP to identify the immediate benefits from demand management programs or identify the current least cost suite of options. Alternatively an LCP approach can be used to design programs that aim to meet designated water conservation targets over time. Australia has a number of examples of such targets, which would be amenable to backcasting. The Victoria Government has recently established a target to reduce Melbourne's per capita water use by 15 per cent by 2010, Sydney Water Corporation has a long standing regulatory requirement to reduce demand per capita by 35% by 2011 from 1991 levels in per capita terms and the ACT Government has targets to reduce per capita demand by 12% and 25% by 2013 and 2023 respectively compared to 2003 levels.

The least-cost evaluation framework

Cost analysis in LCP is distinctive. Firstly, the focus on service provision in LCP allows demand-and supply-side options to be compared on an equivalent basis. The use of levelised cost, calculated in the manner described by Fane *et al* (2002), compares options in terms of their relative unit cost based on the potential to satisfy projected demand. As demand can be satisfied through either increasing the water supplied, or reducing the quantity required, demand management, source substitution, and bulk supply options are treated equally.

Secondly, the concept of cost avoidance is important for LCP (Beecher, 1996). In LCP for water supply, the avoided costs are those costs of supply that are not incurred when water is conserved. Accounting for these costs, which include the direct operating costs of supply, the deferred cost of future augmentation, and the indirect cost savings to consumers (i.e. energy and detergent) are critical in making the case for water conservation.

Finally, LCP for water supply utilises various 'cost tests' which explicitly inform decision-making by and about water utilities (White, 1998). These tests address the 'whole of society' standpoint as well as the financial perspective of stakeholders in supply planning (see Figure 1). The total resource cost and societal cost tests are used in LCP to evaluate the cost effectiveness from a whole of society perspective while the utility and customer cost tests address financial interests (Cal Public Utilities Commission, 2001). For each test, both the costs and benefits (avoided costs) are included.

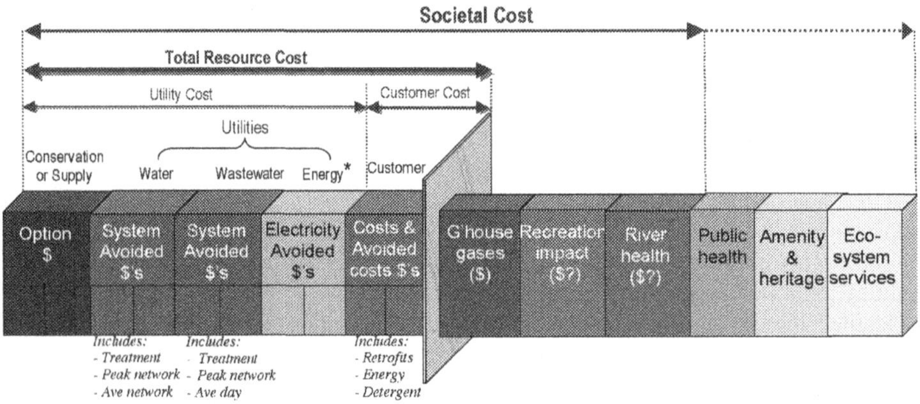

** to avoid double counting, exclude energy costs from Water, Wastewater and customer operating costs.*

Figure 1 Least cost planning cost tests: spectrum of cost and avoided cost

The total resource cost includes all costs and benefits to the utility and its customers. The societal cost test is a better approximation of the whole of society perspective including all those elements in the total resource cost as well as externalities. Transfer payments from governments such as tax credit benefits are excluded. The societal cost is more difficult to estimate because while the economic value of some externalities can be assessed (i.e. greenhouse gas emissions, recreational impacts) a defendable valuation of other impacts of water supply (i.e. river health, public health, heritage) are more difficult or impossible to quantify.

The financial cost tests from the utility and customer group perspectives are used decide how best to roll out programs and to allocate costs fairly amongst stakeholders. They address issues of equity between customer groups and the financial viability of the water utility. This allows the least cost solution to be pursued even when it is not in the water supply utilities current financial interest (Cal Public Utilities Commission, 2001).

Least cost planning for sustainable water management

This section considers the potential of LCP concepts and techniques beyond the design and evaluation of demand management programs.

End-use modelling and the provision of services across the water cycle

The concept of end-use modelling can be extended from it's current application to demand management. In urban water, both wastewater and stormwater infrastructures can be addressed by end-use models. Existing water end-use models can be readily adapted to give estimates of average dry weather sewage flow. As Howe and White (1999) illustrated, end-use models can be used to inform planning for sewer networks and wastewater treatment. For wastewater planning, end-use models could also be enhanced to included pollutant loads. Wastewaters can be conceived as being generated at multiple sources (the end-use services) with each source having various quantity and quality characteristics. Similar to water, aggregate dry weather wastewater flows and pollutant material loads can be derived by combining the end-use or end-source information with data on pollution generated per person or household or from measured concentrations of pollutants present in each stream or source.

End-use models can be combined with stormwater modelling to examine the impact of rain tanks and stormwater capture on both potable demand and stormwater infrastructure requirements. Complete urban water infrastructure scenarios can also be designed around an end-use/end-source models. This has significant advantages over other total water cycle approaches because it focuses scenario design on the actual services to be provided.

For water supply, models which incorporate both daily and seasonal timing for particular end-use demands can also be envisaged. Such an advance would improve projection of peaks demands and facilitate better capacity planning in distribution networks. In some instances, targeted demand management programs could be designed to 'shave' the peaks to within existing system capacities, thus optimising the system. This approach, called 'local integrated resource planning' (Pupp *et al.* 1995) or time and area specific costing (Swisher and Orans, 1995) has already been developed for the electricity sector.

Beyond urban water, end-use models can include agricultural (irrigation demands) and ecological services (environmental flows). Such models could be utilised in catchment management. End-source modelling for water quality planning on a catchment basis is also being developed. In each case an end-use or end-source model will enables water conservation and pollutant source control options to be fully explored and considered equally alongside capital works augmentations.

Backcasting from a sustainable future

As Mitchell and White (2003) state, the development of sustainable water service provision will require the application of both forecasting and backcasting. Backcasting from a sustainable future in relation to an infrastructure system first requires the definition of sustainability goals. At the same time forecasts would be made based on underlying service provision and material flow generation trends. Given the goals and the known trends, interim steps could then be developed. This process is iterative, working backwards to determine the necessary steps towards the desired future, while also working forward based on trend forecasts. By working in this way a number of possible physical and behavioural paths connecting the preferred future with the present could then be developed.

The backcasting concept can readily be extended to address wastewater goals as well as catchment water quality and environmental flow targets. Ideally backcasting would be used to address multiple sustainability goals in a process of planning sustainable development pathways for urban water infrastructures and catchments. Goals might include: water quality targets in catchments, water and nutrient recycling, resource usage, and greenhouse gas intensity.

Backcasting for water systems could also apply deliberative processes in setting sustainability goals. This would ensure that the targets set were both reasonable and acceptable to the community. A deliberative process requires a forum such as citizen juries, consensus conferences, or planning cells. Such techniques assess community opinions and act to educate these opinions through briefing materials (Carson and Gelber, 2001). Critically these techniques also allow citizens to discuss and deliberate on the issues in question.

Challenging the assumption that future urban water systems will follow the conventional model of centralised networks and supply of one quality of water, backcasting can be used to generate alternative scenarios. Alternatives might be based on distributed wastewater and stormwater systems as well as utilise distributed source of supply (such as neighbourhood run-off and the local reuse of effluent). Such scenarios could then be compared to more conventional systems that might also provide future service demand and meet the sustainability goals set.

Extending the least-cost framework

As with LCP for water supply, designing options around an end-use demand projection model can change the relative costs of options for wastewater, stormwater and whole infrastructure scenarios. The concept of levelised cost, which in water supply LCP is the unit cost of water conserved or supplied, can be applied to both wastewater and stormwater volumes. Building alternative infrastructure scenarios for water in a new urban development around an end-use model of forecast demand also has advantages at both the scenario design and life cycle costing stages of an evaluation.

The concepts of avoided costs and the utility of various 'cost tests' can likewise be extended and applied to infrastructure planning across the water cycle. For example, cost tests which consider the public sector as a whole and the developer's perspective would be of benefit for State Government planning of new development areas. As Figure 1 illustrated, the spectrum of avoided costs for an option also extends beyond those externalities which are simple to value and include in the societal cost. Moving along the spectrum, more difficult to value externalities, such river health, are followed by impacts where reasonable and non-contentious monetary valuation is not possible (such as public health or ecological services).

As O'Conner (2000) contends, not all impacts can be translated into dollar term. A frontier of monetisation exists beyond which monetary evaluations are not meaningful. Crossing the frontier people can not consider impacts in a solely economic manner and information on impacts needs to be organized in a non-monetary form (O'Connor, 2000). At this point we argue that further aspects must then be accounted for qualitatively. This could be as an absolute constraint on the options considered or as sustainability goals for backcasting. Such an approach would treat critical criteria for water system sustainability as limits on the options or goals for scenario development. Less critical criteria (although not necessarily less important) would be treated as costed externalities within the societal cost test.

Conclusions

Least cost planning is based on the key idea that demand for water is in fact derived from the demand for the services that water provides. Least cost planning concerns itself with how water systems mediate resource usage and environmentally impact in producing these water related-services for system users. In LCP, options and scenarios are constructed around an end-use model. This exemplifies the focus on service provision. Importantly, end-use models allow demand management, source substitution and pollutant source control measures to be full considered.

In LCP forecasting and backcasting are complimentary. When LCP is used in a backcasting study, various means of moving from where we are to the determined desirable future state where we want to be are compared. The LCP approach permits this exploration of alternative pathways only because of the detailed forecasting based on underlying service provision trend embodied in the end-use model.

Options evaluation in LCP is characteristic because of the focus on service provision in modelling and analysis. The evaluation framework in LCP also moves beyond cost benefit analysis. An extended LCP framework can account for the qualitative aspects of sustainability together with cost analysis. It provides alternative means for incorporating externalities and sustainability criteria together with cost.

An LCP evaluation framework also accounts for the various financial perspectives of stakeholders. This means cost sharing arrangements can be considered based on the partitioning of costs between parties relative to the benefits. The LCP framework thereby allow an exploration of which stakeholders will benefit and how much it will cost to move along the different pathways towards sustainability.

In conclusion, because of its various distinctive and beneficial attributes it is our contention that LCP concepts and techniques have significant potential for use in assessing sustainable outcomes across the water cycle.

References

Beecher, J. (1996) Integrated resource planning for water utilities. *Water Resources Update*, issue 104.

California's Public Utilities Commission. (2001) California standard practice manual: economic analysis of demand-side programs and projects.

Dreborg, K (1996) Essence of backcasting. *Futures* **28** (9) 813–828.

Fane S, J. Robinson, and S. White, (2003), The use of levelised cost in comparing supply and demand side options. *Wat. Sci. Tech: Wat. Supply* **3** (3).

Howe, C. and White, S. (1999). Integrated resource planning for water and wastewater: Sydney case studies. *Water International*, **24** (4), 356–362.

Mitchell, C. and White, S. (2003) Forecasting and backcasting for sustainable urban water futures *Water* **30** (5).

O'Conner, M. (2000) Natural capital, policy research brief no.3, the environmental valuation in Europe project.

Pupp, R, Woo, C., Orans, R., Horii, B. and Heffenerp, G. (1995) Load research and integrated local T&D planning. *Energy* **20** (2), 89–94.

Robinson, J. B. (1982) Energy backcasting: a proposed method of policy analysis. *Energy Policy* **10** (4), 337–344.

Swisher, J. and Orans, R. (1995) The use of area-specific utility costs to target intensive DSM campaigns. *Utilities Policy*, **5** (3/4), 185–197.

White S., and Fane, S. (2002) Designing cost effective water demand management programs in Australia. *Wat. Sci. Tech.* **46** (6).

White, S. (1998) *Wise water management : a demand management manual for water authorities.* Water Services Association of Australia.

The use of sustainability criteria for the selection and comparison of sanitation systems

P. Bracken (GTZ)*, E. Kvarnström (EcoSanRes), A. Ysunza (Ceciproc), E. Kärrman (Urban Water), A. Finnson (Stockholm Water), and D. Saywell (WSSCC)**

*GTZ Ecosan Team, Dag Hammarskjöld Weg 1-5, 657 26 Eschborn, Germany
(E-mail: Patrick.Bracken@gtz.de)
**Programme Advisory Committee of EcoSanRes, VERNA Ecology, Inc. Malmgårdsvägen 14, SE-116 38 Stockholm, Sweden (E-mail: Elisabeth@verna.se)

Abstract Given the current global sanitary crisis and the measures proposed to address this, particularly the plan of action to reach the Millennium Development Goals, there is a desperate need to provide sanitary systems which can satisfy existing needs, whilst ensuring that this is done in the most sustainable manner possible. There is therefore a need to integrate the more holistic considerations of sustainability into current decision making processes for sanitation, from the micro to the macro level. This can be achieved if a careful identification of criteria indicating the sustainability of the system is carried out and these criteria are used as a basis for decision making. The objective of this paper is to present a broad list of sustainability criteria for sanitation developed in a series of two international and inter-organisational workshops (Eschborn, Germany, December 2003 and Stockholm, Sweden, April 2004). This list should inspire decision-makers, donor agencies, NGO's, and CBO's to assess the sustainability of different sanitation alternatives before making sanitation decisions and implementing programmes and projects. It also aims to make them aware that establishing and using criteria for sustainability need not be part of a purely academic task, and can serve to provide an extremely well founded basis for discussion, increasing the participation opportunities for all stakeholders in a decision making process.

Keywords Decision making criteria; sustainability and sanitation

Method

Introduction

Recent estimates state that world-wide at least 2.6 billion people lack access to improved sanitation (UNICEF and WHO, 2004), defined by the WHO as being a connection to a public sewer or septic tank system, a pour-flush latrine, a simple pit latrine or a ventilated improved pit latrine (WHO, 2000). This global sanitation crisis has been recognised by the international community and a concrete Millennium Development Goal has been set to halve the number of people without access to adequate sanitation by 2015 (UN, 2002).

Currently, legislation and decision making procedures for choosing sanitation systems appear to be determined by norms, standards and practices that have developed within the framework of conventional, end-of-pipe sanitary thinking, based mainly on initial investment and operation and maintenance cost of the system. However, if our objective is to provide sustainable sanitation systems, a more integrated view of sanitary provision must be adopted where the search for sanitary solutions is geared towards finding sustainable systems not confined by the boundaries defined economically by initial investment and operation and maintenance costs, environmentally by limiting effluent concentrations and technically by systems that have proven to be unsustainable.

One way of guiding decision making processes towards choosing socially, economically and environmentally sustainable sanitation is to use sustainability oriented criteria in decision making processes for the comparison and selection of sanitary systems. Such sustainability oriented criteria based approaches can, and we believe should, be used across the entire range of planning and implementation levels – from the macro level, by including them in the terms of reference for sanitation projects financed on a national or international level, to the micro level where they can be extremely useful when used by a community to select their sanitary system according to their needs and vision of sustainability. Developing and using a context-specific list of criteria to indicate the overall sustainability of a sanitation system therefore helps gear the decision making process towards the issues relevant to the different stakeholders, and away from basic economic and techno-centric discussions. This allows more room for the implementation of innovative sanitation solutions that are tailored to the needs of the system users (Tischner and Schmidt-Bleek, 1993).

The use of criteria in order to assess sustainability of sanitation systems has been suggested in the literature. Larsen and Gujer (1997) underlined the need to focus on functions that the urban water management system shall provide in order to be sustainable and computer-based multi-criteria analysis tools for assessment of sanitation systems have been proposed by different actors, including the Swedish water and sanitation research group Urban Water (http://www.urbanwater.org/default_eng.htm) and the Dutch researcher van der Vleuten-Balkema (2003).

However, despite this work, the use *and* usefulness of sustainability criteria for sanitation decision making remains largely unrecognised in real-life project planning and implementation. This may be due to the fact that much of this work has so far been mainly the domain of a relatively small circle of academics. The use of abstract theories and computer models may actually be serving to discourage practitioners from using sustainability criteria in their decision making processes.

Method

Setting the scene

In literature, many references to sustainable sanitation can be found without any accompanying definition of what this term might actually imply. As work aims to identify criteria that can be used to assess the sustainability of sanitary systems it was considered important to at least arrive at a definition of sustainability with regard to sanitation that we as a group could use as a basis for our work. However, in order to do this, clear boundaries were needed to define what a sanitation system is. Without this clear definition of what the boundaries of a sanitation system are, a comprehensive comparison of different sanitation systems was seen to be impossible. It was therefore agreed that a sanitation system comprises the users of all parts of the system, along with the collection, transport, treatment of human excreta, solid waste, industrial wastewater, and storm water, and the management of the resulting end products.

This definition explicitly recognises that sanitation can not simply be seen as being an element contained entirely within the water cycle, even though conventional water based sanitation is integrated completely into it, and here the most obvious impacts of poor sanitation can be easily observed. These boundary conditions also deliberately include the social aspect of sanitation, the economic and logistical side, and the idea of resource management.

Taking these boundary conditions and combining them with an understanding of the Brundtland definition of sustainable development, as development that meets the needs of current generations without compromising the ability of future generation to meet their own needs, we have based our range of criteria on the understanding that a sanitation system that is sustainable protects and promotes human health, does not contribute to environmental degradation or depletion of the resource base, is technically and institutionally appropriate, economically viable and socially acceptable.

The definition of sustainability with regard to sanitation was however still seen to be quite abstract and too general to base a series of criteria upon. A series of practice oriented principles were therefore sought and used to help bridge the gap between theory and practice. For this purpose the four 'Bellagio Principles' were used. These principles were the conclusion of the work of a group of experts from a wide range of international organisations involved in environmental sanitation that met in February 2000 in Bellagio, Italy. The group called for a radical overhaul of conventional sanitation policies and practices world-wide and formulated the four principles which have formed the basis of the Household Centred Environmental Sanitation approach. They were endorsed by the members of the WSSCC at their 5th Global Forum in 2000. The Bellagio Principles are:

1) Human dignity, quality of life and environmental security at household level should be at the centre of the new approach (to sanitation), which should be responsive and accountable to needs and demands in the local and national setting:

2) In line with good governance principles, decision-making should involve participation of all stakeholders, especially the consumers and providers of services

3) Waste should be considered a resource, and its management should be holistic and form part of integrated water resources, nutrient flows and waste management processes.

4) The domain in which environmental sanitation problems are resolved should be kept to the minimum practicable size (household, community, town, district, catchment, and city) and wastes diluted as little as possible.

Whilst providing a practical guide to identifying sustainability criteria, the principles were not followed dogmatically by our working group, particularly when it was felt that they did not directly address a certain concern or aspect of sanitation.

Developing the criteria

The use of criteria today

The use of criteria in order to assess the sustainability of sanitation systems is certainly not new. Larsen and Gujer (1997) underlined the need to focus on the functions the urban water management system should provide to be sustainable. These functions are (1) to guarantee urban hygiene, (2) to assure drinking water of good quality and enough quantities to allow for personal hygiene, (3) to prevent flooding/allow for drainage of urban areas, (4) to integrate urban agriculture into urban water management, and (5) to provide water for pleasure and for recreational aspects of urban culture. In addition to this, Larsen and Gujer (1997) underlined the need to carefully identify the system boundaries when looking for sustainable solutions in order to avoid exporting problems in either time or space – the inter and intra-generational aspects of sustainability. Various numeric tools have been proposed to assess sanitation system sustainability using a range of multiple criteria analysis techniques. One example is a toolbox, currently being developed by the Swedish water and sanitation research group Urban Water. The toolbox will include models for substance flow analysis, microbial and chemical risk assessment, cost estimates, user aspects and more (http://www.urbanwater.org/default_eng.htm). In 2003, van der Vleuten-Balkema also presented a computer model for the assessment of domestic water systems, which included sustainability indicators grouped under the categories: functional, economic, environmental and socio-cultural.

The usefulness of criteria in the planning process

Despite the reports on criteria for sustainable sanitation referred to here, and others not mentioned, we feel there is a lack of recognition and use of criteria for sustainable sanitation in project planning. The use of criteria for sustainable sanitation would be a great help in making informed decisions for investment banks and municipal, regional or national authorities when planning for sanitation provision. They would equally serve very well when planning for sanitation provision to peri-urban areas, where one crucial question is often whether more sustainable sanitation systems would be provided by on-site solutions rather than through a connection to a central wastewater treatment plant. Sustainability criteria can also be used for external evaluation of sanitation system compliance to sustainability, and also internally in an organization or company to evaluate existing sanitation systems with regard to their sustainability (Lundin *et al.*, 1999).

Results

Suggestion of criteria for assessment of sustainability of sanitation systems

The results of our deliberations, a list of criteria that might be of importance in the assessment of the sustainability of sanitation systems, can be seen in Table 1. This list has been developed from criteria/functions/indicators outlined in the literature (van der Vleuten-Balkema, 2003; Hellström *et al.*, 2000; Larsen and Gujer, 1997; Larsen and Lienert, 2003; Lennartsson, 2004, Urban Water, 2004) and amended, adjusted and supplemented during the workshops organised by the German Development Co-operation GTZ in Eschborn, 2003, and by the research and development network for ecological sanitation, EcoSanRes, in Stockholm, 2004. As can be seen, the criteria themselves differ from the conventional triple bottom line usually accepted as the three pillars of sustainability – economy, society and the environment. We identified five main categories; health, environment, economy, socio-culture, and the technical function, which we believe to be of relevance when considering sustainability for sanitation systems. The protection of human health was and is the main aim of providing sanitation. To stress this fact, health has been taken as a category in itself and not as an element of environmental or social considerations. The technical function of the system was also considered an important criteria for sustainable operation, and this could not be satisfactorily addressed under the triple bottom line. The categories are described below in more detail:

Health

The *entire* sanitary system should be hygienically safe, posing as small a risk as possible to human health. This covers the use of the sanitary installation, collection, transport, treatment and end destination of the treated products. The risk of infection from leaking sewers or pit latrines to the groundwater, for example, should also be included, as should the risk of being infected when bathing in lakes or the sea near overflows or treatment plant outlets.

Environment

In many industrialised countries sanitary systems have been further developed to protect the environment against possible detrimental effects of treated effluent. When considering the environmental sustainability both emissions to different recipients (water, soil, and air), as well as resource use by different sanitation systems during the construction and operation phase must be accounted for. Moreover it is important to consider the quality of the treatment product for possible reuse in agriculture.

Economy

The capacity to pay for sanitation is clearly a decisive factor in choosing the system. Among the users, the capacity to pay is an important criteria for sustainability. However in the end it may be their willingness to pay that will define within what range the costs (both for construction and O&M) can vary and services be sustained financially by the population. In the discussions of the working group some time was spent considering to what degree the other criteria may eventually be reducible to purely economic considerations. Considerations such as the impact on local development (e.g. local private sector), health benefits etc. were seen as possibly being expressible in monetary terms. However, this reduction of criteria to their estimated economic value was seen as not being sufficient to allow a thorough consideration of all of the aspects.

Socio-culture

Clearly the objectives of sanitation are to protect human health and the environment, however, sustainability in sanitation cannot be based only on these objectives. There is an obvious need to include social criteria as they are crucial to the sustainability of the use and the services provided by the system. It is possible to distinguish at least three very different types of important criteria within this category, namely cultural acceptance, institutional requirements, and perceptions of sanitation. Society is more dynamic and flexible than criteria such as human health and the environment and therefore the socio-cultural criteria are subject to a more dynamic change with time than human health and the environmental criteria. How they system is perceived, legislated for and accepted is therefore changeable with time. Although improving human health and protecting the environment are the main objectives for planners and politicians, this might not be enough to sell the sanitation concept to future users. It is also important to recognize that the prime driver for sanitation might be security and status rather than health and environment as recent studies have shown (Holden, 2003; Wirbelauer *et al.* 2003). Another sanitation driver could be the possibility of increased food security if the sanitation solution can provide hygienically safe fertilisers.

Table 1 Suggested list of criteria to assess the sustainability of sanitation systems

CRITERIA	INDICATOR
HEALTH	
Risk of infection of complete use of system	Risk assessment or qualitative
Risk of exposure to hazardous substances: heavy metals, medical residues, organic compounds	Risk assessment or qualitative
ENVIRONMENT	
Use of natural resources, construction and O&M	
Land (investment, constr. and O&M)	m^2/pe
Energy (constr. and O&M)	MJ/pe
Construction material (constr.)	Type and volume
Chemicals (constr. and O&M)	Type and volume
Fresh water (O&M)	
Discharge to water bodies	
BOD/COD	g/pe/yr
Impact on eutrophication	g/pe/yr of N and P
Hazardous substances: heavy metals, persistent organic compounds, antibiotics/medical residues, hormones	mg/pe/yr
Air emissions	
Contribution to global warming	kg of CO_2 equivalent/yr
Odour	Qualitative
Resources recovered	
Nutrients	% of incoming to the system
Energy	% of the consumption within the system
Organic material	% of incoming to the system
Water	% of incoming to the system
Quality of recycled product (released to soil)	
Hazardous substances: heavy metals, persistent organic compounds, antibiotics/medical residues, hormones	mg/unit
ECONOMY	
Annual costs, including capital and maintenance costs	Cost/pe/yr
Capacity to pay – user (% of available income)	Disposable income/pe
Local development	Qualitative
SOCIO-CULTURAL (Institutional and User)	
Willingness to pay (% of available income)	Reasonable % of income – defined by users
Convenience (comfort, personal security, smell, noise, attractiveness, adaptability to different age, gender, and income groups)	Qualitative
Institutional requirements	Qualitative
Responsibility distribution	Definition of level of organisation
Current legal acceptability	Qualitative
Appropriateness to current local cultural context (acceptable to use and maintain)	Qualitative
System perception (complexity, compatibility, observability – including aspects of reuse)	Qualitative
Ability to address awareness and information needs	Qualitative
TECHNICAL FUNCTION	
System robustness: risk of failure, effect of failure, structural stability, robustness against extreme conditions	Qualitative
Robustness of use of system: shock loads, abuse of system	Qualitative
Possibility to use local competence for construction and O&M	Qualitative
Ease of system monitoring	Qualitative
Durability / Lifetime	Qualitative
Complexity of construction and O&M	Qualitative
Compatibility with existing systems	Qualitative
Flexibility / adaptability (to user needs and existing environmental conditions – high groundwater level, geology etc.)	Qualitative

Technical function

The technical reliability and performance of the system is key to its success and ultimately its sustainability. One of the more important sub-criteria under this category is probably robustness, both within the system (to be able to receive varying loads) and externally (to be able to withstand varying extreme environmental conditions as well as user abuse).

The technical functioning of the system is seen as perhaps the most flexible group of criteria. In this regard, we would strongly promote the logic behind identifying and then adapting technologies to the needs of the situation rather than trying to make a situation comply with the demands of a favoured technology. It is also much easier to adapt a technology to what people need than to adapt people to what a technology requires.

For the suggested list of criteria we would like to emphasise that we are well aware that one cannot prepare a general list of sustainability criteria that will be universally applicable in all contexts. The requirements of sustainability are dictated by context, and can change with time. The intention of this list is rather to inspire sanitation planners on all levels to include the concept of sustainability criteria in their planning processes, and to actually define, in a participatory manner with the relevant stakeholders, what criteria their planned sanitation system should fulfil in order to be sustainable in their context. The list presented here should provide certain 'core' considerations of sustainability, which should be considered as a minimum in any basic list to assess a sanitation system.

One way in which the criteria may be used would be as the focal point of a multiple criteria decision making process. There is a broad range of techniques that can be employed for this. This need not be an extremely complicated process and methods are available that could facilitate this when working with grass roots level groups. In such cases it is often considered appropriate to allot agreed weights to the criteria identified, relative to their importance to the stakeholders, and to use these weighted versions when deciding. Each stakeholder group may have criteria which are of particular importance to them, and different criteria may be of greater relevance at different times in the decision making process.

Conclusions and recommendations

The list of criteria presented in this submission is seen as a contribution to advancing the discussion regarding the use of criteria in sanitation decision making. It is based on the different criteria / functions / indicators already used by those working in this field, which have been discussed and adapted during the course of two international workshops.

The list has been divided into five main categories: health, environment, economy, society and culture, and technical function of the system, of which health and technical function are not part of the commonly used triple bottom line definition of sustainability. The reason for this is that one of the main functions of sanitation systems is to protect human health. Moreover, for any sanitation solution to be sustainable it is extremely important to consider its technical function.

The criteria in the list have been deliberately drawn up in quite a broad and general fashion in recognition of the fact that such criteria need to be developed 'in situ' for them to be of real relevance in decision making. The general list provided here could be used as a basis for this and adapted to suit the needs of the situation with the participation of all stakeholders. The list of criteria presented here can be used to narrow down and focus discussions among decision-makers and also inspire the development of context-specific sustainability criteria in the actual planning situation.

We strongly recommend the use of sustainability criteria in any strategic sanitation planning and decision-making process whether on a macro or micro project level. Moreover, sanitation sustainability criteria can be used for follow-up and evaluation of sanitation systems. They should be integrated into current, commonly used decision making processes as well as into more recently introduced approaches to sanitation, such as the guidelines for implementing the Bellagio Principles in urban environmental sanitation services from the WSSCC. On the macro level, international financing and donor organisations must be encouraged to make sanitation decisions on the basis of sustainability considerations. To this end approaches have recently been made to the UN Millennium Task Force 7 on water and sanitation, and to the World Health Organisation to include sustainability criteria in upcoming publications. We would greatly welcome any feedback to amend and improve the criteria list presented here and any additional input to improve this work.

Acknowledgements

This work was jointly financed by the Swedish International Development Co-operation Agency, Sida, through funding of the research and development network for ecological sanitation, EcoSanRes and by the German Development Co-operation, GTZ.

References

World Health Organization and United Nations Children's Fund (2000) Global water supply and sanitation assessment: 2000 Report.

Hellström, D., Jeppsson, U. and Kärrman, E. (2000) Systems analysis of urban water management. *Environmental Impact Assessment Review* **20** (3).

Holden, R., Terreblanche, R. and Muller, M. (2003) Factors which have influenced the acceptance of ecosan in South Africa and development of a marketing strategy. In *Proceedings of the 2nd International Symposium on ecological sanitation*, pp. 167–174.

Larsen, T. and Gujer, W. (1997) The concept of sustainable urban water management. *Wat. Sci. Tech.* **35** (9), 3–10.

Larsen, T. and Lienert, J. (2003). Societal implications of re-engineering the toilet. *Water Intelligence Online*, March.

Lennartsson, M. (2004). Review of alternative sanitation systems. EcoSanRes working paper (draft version).

Schertenleib, R. *et al.* (2003) Guidelines for the implementation of the Bellagio Principles and the Household Centred Environmental Sanitation Approach (HCES). In *Proceedings of the 2nd International Symposium on Ecological Sanitation.*

Tischner, U. and Schmidt-Bleek, F. (1993) Designing goods with MIPS. *Fresenius Envir. Bull.* **2**, 479–484.

United Nations (2003) Water for people, water for life. The United Nations World Water Development Report.

UNICEF and WHO (2004). Meeting the MDG water and sanitation target: a mid-term assessment of progress.

Urban Water (2004) MIKA – Methodologies for Integration of Knowledge Areas. Surahammar Case Study (draft in Swedish).

van der Vleuten-Balkema, A. (2003) Sustainable wastewater treatment – developing a methodology and selecting promising systems.

Wirbelauer, C., Breslin, E. D., Guzha, E. (2003) Lessons learnt on ecosan in Southern Africa – towards closed-loop sanitation. In *Proceedings of the 2nd International Symposium on ecological sanitation*, pp. 135–142.

Selection of indicators for sustainable management of an urban stormwater system

L. J. Cobiac

*Cooperative Research Centre for Coastal Zone, Estuary and Waterway Management and Department of Mathematics, The University of Queensland, St Lucia, 4072, Australia

Abstract Indicators are a popular method of measuring sustainability at the global or national scale as a means of directing policy development, and they are increasingly being used to implement sustainability at the planning level. In this paper, we present a process of indicator selection for management of an urban stormwater system, and evaluate the indicators for a simple management scenario. Indicator selection and assessment is commonly approached from the Triple Bottom Line concept of sustainability, where sustainability is equated to simultaneous achievement of social, environmental and economic goals. But some have questioned this concept of sustainability because it contributes to beliefs that the well-being of society and the environment are independent, and usually in opposition. Thus in this paper we have taken an alternative approach, which is based on deriving sustainability objectives and measurement criteria from the fundamental principles of sustainability. A key advantage of this principles-based approach was that it places greater emphasis on the human behavioural changes that are fundamental to achieving a state of sustainability. We concluded that this approach to indicator selection is vital to the way that indicators will be aggregated or compared in the sustainability assessment, and hence the influence of the information on planning decisions.

Keywords Indicators; sustainability assessment; urban stormwater

Introduction

Sustainability is a concept that is intuitively understood, but a challenge to define in precise terms, and even more challenging to put into practice. To incorporate sustainability into planning strategies, we need ways of assessing the sustainability of our management actions so that we can measure progress towards sustainability goals.

Indicators are commonly used as a measure of sustainability at the global or national scale as a means of directing policy development. State of the Environment reporting (e.g. Australia State of the Environment 2001 (Australian State of the Environment Committee, 2001)) is a typical example. But indicators can also be used to guide action at the ground level.

In water systems planning indicators have been used for many years, providing information on environmental impacts and technical issues, to assist with decision-making. But with growing recognition of the need for sustainable use of water resources and disposal of wastewaters to the environment, the selection of indicator is increasingly being extended to incorporate social and economic information into the planning and decision-making process. Balkema *et al.* (2002), Walmsley (2002) Hellstrom *et al.* (2000), Lundin *et al.* (1999), and Ashley *et al.* (1999) are all recent examples addressing sustainability indicator assessment of urban water systems or catchments. None, however, have focused on the development of indicators to assess sustainability in managing stormwater systems.

We are developing a system to assist urban planners with sustainability assessment of stormwater management scenarios (see Figure 1). The system consists of two parts, a sustainability framework of indicators, and a fuzzy rule-based system for the integration of indicators in the assessment. This paper focuses on the development of the sustainability framework and indicators, with evaluation of the indicators for a simple management scenario.

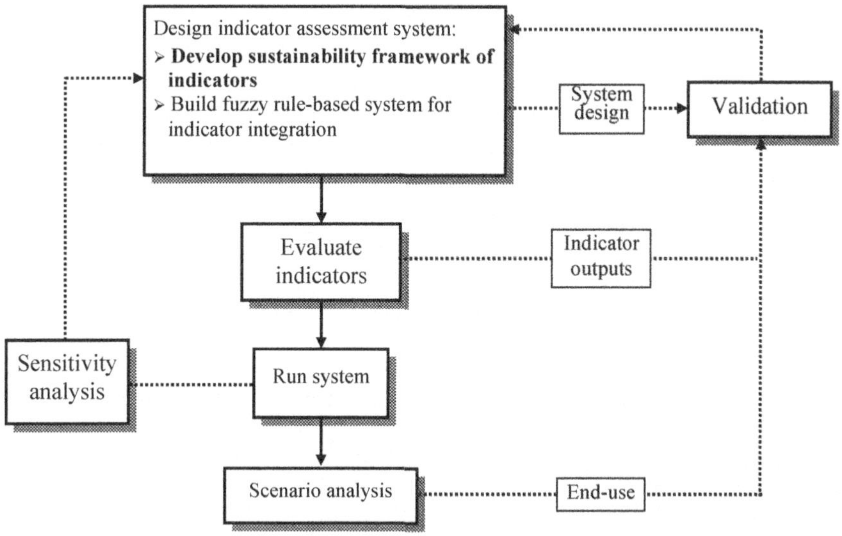

Figure 1 The overall development and testing process for the sustainability assessment method. The development of the sustainability framework and indicators is the focus of this paper

The indicator selection process

There are many approaches to selecting a set of indicators; choice of method depends on the purpose of the assessment. If the assessment is for comparison of different locations (e.g. as a policy tool for 'rating' the performance of different catchments) then choosing from a standard set of indictors is important. The set of indicators for State of the Environment reporting (see ANZECC State of the Environment Reporting Task Force (2000)) are an example of this. If, however, the assessment is for planning purposes, as in this case, where it is a means of comparing management scenarios for the one catchment, then indicators should be specific to the catchment and the system being evaluated. The indicators should provide information that reflects local interpretations of sustainability.

A sustainability framework represents an interpretation of sustainability, and provides the basis for selecting indicators. Framework development typically begins with division of system sustainability issues along traditional disciplinary lines (e.g. environmental, social and economic). Balkema *et al.* (2002), for example, grouped indicators into economical, environmental, technical and social-cultural categories. Hellstrom *et al.* (2000) derived five categories: health and hygiene, social-cultural, environmental, economic, and functional and technical. While Ashley *et al.* (1999) derived three categories: economic, ecological and socio-political. The specific terminology and number of categories varies, but they have in common an approach to interpreting sustainability that is akin to the Triple Bottom Line approach, where sustainability is interpreted as the simultaneous achievement of social, economic and environmental goals.

The Triple Bottom Line concept of sustainability has received criticism from some quarters. Pope (2003), for example, has argued that a balancing of environmental, social and economic objectives will not necessarily equate to a sustainable solution, and recommends an alternative approach based on the fundamental principles of sustainability. This begins with definition of sustainability principles, based on interpretation of the fundamental principles of sustainability in the context of both the system being assessed and the wants and needs of those who will be involved in using or managing the system, and is followed by definition of criteria against which sustainability will be measured. This alternative approach could lessen the perception that the well-being of society and the environment are independent, and are usually in opposition, a misconception that is reinforced by the Triple Bottom Line concept. It would also more directly address the human behavioural changes that are fundamental to achieving a state of sustainability.

In this study we have adopted this principles-based approach to developing the framework for which the indicators are selected. The objectives and criteria form the foundation of the framework. This process is illustrated in Figure 2.

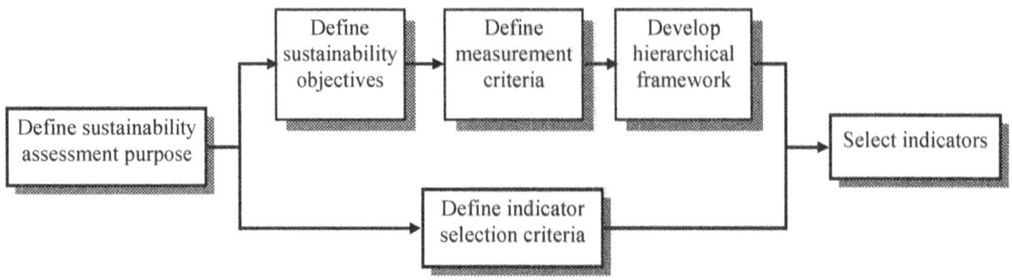

Figure 2 The process of developing the sustainability framework and selecting indicators

Developing the sustainability framework

The first step in developing the framework is to define the sustainability objectives. Urban stormwater systems are designed to deal with the problems created by an unsustainable lifestyle (Butler and Parkinson, 1997). They are needed to protect the health and safety of the community, and prevent property damage and disruption from flooding, when the environment's capacity to assimilate wastes is exceeded. To be sustainable, the management plan should move the system towards a state that does not exceed the natural environmental capacity to produce resources and assimilate wastes. It is also vital that it addresses the means and motivation for people to maintain and continue driving the system towards this ideal state of sustainability. Thus there are two key objectives for stormwater management in urban catchments. The first objective is to achieve a balance between human use of freshwater resources and disposal of polluted stormwater wastes, and the environment's capacity to provide resources and assimilate the wastes. And the second objective is to maximise the potential for either maintaining the state of balance if it is attained or achieving progress in the sustainable direction.

The next step in developing the framework involves defining criteria for assessing the extent to which a stormwater management plan will meet the objectives that have been identified. The key criteria include a healthy ecosystem, meeting people's basic needs for health and safety, the availability of sufficient means of implementing the management plan (including finance, expertise, labour and resources), and providing individuals with the means and the motivation to carry out on-going management responsibilities. These are further broken down into sub-criteria. This hierarchy of objectives, criteria and sub-criteria forms the framework of illustrated in Figure 3.

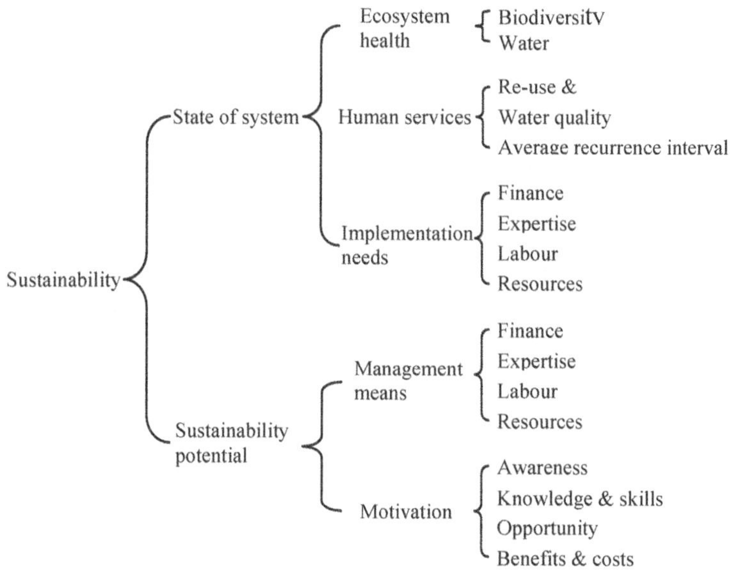

Figure 3 The hierarchy of sustainability indices forming the framework

Indicator selection and evaluation

There are many criteria that have been suggested for guiding selection of the indicators. These include *representative, reliable, feasible* (OECD, 2002), *simple, quantifiable, communicable* (Walmsley, 2002), *clearly and unambiguously defined, qualitative or quantitative, with a hierarchy of specificity* (Bossel, 1996), *sensitive to change, predictive, available reference/threshold values, ability to measure reversibility or controllability, appropriate data transformation, integrative ability* and *relative ease of collection and use* (Liverman *et al.*, 1988). Some of these criteria could overly restrict the selection of indicators and thus lead to a less comprehensive assessment of sustainability. A limitation, for example, in some indicator assessments (e.g. Lundin *et al.* (1999)) is that selection of indicators is restricted to those that can be quantified and for which data are available. This excludes criteria, particularly social criteria, from the analysis.

In this study, there is no need to restrict indicator selection to those that can be quantified. In the second part of the sustainability assessment tool, the indicators presented in this paper will be integrated using a fuzzy rule-based system. The advantages of this fuzzy rule-based approach are that it uses linguistics, so input to the system can be either quantitative or qualitative, and it can handle imprecise inputs. This ensures that all criteria can be included in the assessment, even if they cannot be measured numerically or defined precisely. Thus, the selection of indicators in this case, was based on the ability of the indicators, when combined, to provide an adequate representation of the variable, and the ready availability of information or expertise for evaluation. The need to be quantifiable and precisely definable was not necessary.

Based on these criteria, indicators were selected for each variable at the lowest level of the framework (e.g. biodiversity, water quality, etc.). The indicators were selected to provide information about both the current state of each variable and the likely impact on that variable of a management scenario.

The evaluation of indicators has initially been carried out for a simple management scenario involving the installation of a sedimentation basin in a small urban catchment in South-East Queensland. Information for evaluation was obtained from a variety of sources, including data (e.g. water quality), modelling (e.g. flooding potential), local expertise and relevant literature (e.g. analyses of the performance of stormwater management methods). Given the imprecision of some variables, they were defined on a scale of *low*, *medium* and *high*, with the option of qualifiers *somewhat* and *very*, which is suitable for input to the fuzzy rule-based system. The evaluation of indicators is summarised in Table 4.

Discussion

The validation of results from an indicator assessment differs from the validation process for a predictive model; the qualitative nature of many indicators prohibits the traditional method of comparison with measured data. Bockstaller and Girardin (2003) have proposed a three-part validation process for indicators, involving validation of indicator design (Is it scientifically founded?), indicator output (Is the output realistic?), and the end-use (Is it useful as a decision aid?). Expert judgement and peer review play a key role in this process. The presentation of the indicator selection method and the evaluation of indicators in this paper is, in effect, the first stage of Bockstaller and Girardin's validation process.

A primary step in the indicator selection process is the development of the sustainability framework. In this research, the framework was developed using a principles-based approach. OECD (OECD, 2002) recommends that the framework should be in a form that is easily communicated, hierarchical, logical and representative of the system being assessed, and this framework meets these criteria.

This framework structure also has a number of potential advantages. In particular, it emphasises the need for management planning to address the means and motivation required for long-term management of stormwater systems and the on-going pursuit of sustainability goals. This alternative structure also discourages the common perception that sustainability will only be achieved through a trade-off between human and ecosystem needs, implying that a sacrifice in human quality of life is required. This is a negative and potentially harmful interpretation of sustainability that should not be reinforced.

In theory, the specific structure of the framework may have little impact on the actual set of indicators that are selected. Yet its structure is arguably the most important part of the indicator selection process because it dictates the organisation and presentation of the indicator information to planners and decision-makers. The structure of the framework affects the way in which indicators will be aggregated or compared in the sustainability assessment, and hence the influence of this information on planning decisions. Thus, this principles-based approach to indicator selection is critical to the outcome of the sustainability assessment process.

Table 1 The sustainability indicators

Sustainability Variable	Indicators	Current State[1]	Management Scenario[2]
Ecosystem health			
Biodiversity	Proportion of streamline with native riparian vegetation	Low	Very low
Water quality	DO, pH, Turbidity, TP, TN, SS, BOD relative to regional guidelines	Low	Medium
Human services			
Re-use and recycling	Water storages used for re-use or recycling	Somewhat medium	Very low
Water quality	FC relative to regional guidelines	Somewhat medium	Low
ARI	10 year, 100 year ARI hazard category	Medium	Somewhat medium
Implementation needs			
Finance	Finance required for implementing management	Medium	Medium
Expertise	Expertise required for implementing management	Low	Medium
Labour	Labour required for implementing management	Low	Medium
Resources	Resources required for implementing management	Low	Medium
Management means			
Finance	Ability to meet finance needs for on-going management	Medium	Somewhat high
Expertise	Ability to meet needs for expertise for on-going management	Low	Medium
Labour	Ability to meet labour needs for on-going management	Low	Somewhat high
Resources	Ability to provide resources for on-going management	Low	Medium
Motivation			
Awareness	Recreation facilities, educational awareness campaigns, aesthetic impact	Somewhat low	Very low
Knowledge & skills	Educational activities	Low	Very low
Opportunity	Community/catchment group facilitation	Somewhat low	Very low
Benefits and costs	Targeted incentives to participate (e.g. rebates)	Low	Very low

[1]Information and data provided by Caloundra City Council; Connell Wagner (2001). [2]Information provided by John Wilson and Partners (2000); MUSIC modelling; Engineers Australia (2003), Auckland Regional Council (2003), EPA Victoria (1999), NSW EPA (1997), Semple and Bycroft (1996), and US EPA (1993).

Conclusion

In this paper, we have presented a process of indicator selection for stormwater management. This process consisted of a series of steps, beginning with clarification of the purpose of the sustainability assessment, followed by a series of steps to develop a sustainability framework, identification of indicator selection criteria, and finally, the selection of the indicator set. These indicators were then evaluated for a simple management scenario.

The key step in this process is the development of the framework, which dictates the organisation and presentation of the indicator information to planners and decision-makers. The framework is commonly structured around the environmental, economic, social and sometimes other aspects of the system. But in this study, we took and alternative approach that was based on a series of sustainability objectives and measurement criteria.

In assessing management scenarios, this principles-based approach is essential, because it affects the way in which indicators will be aggregated or compared in the sustainability assessment, and hence affects the influence of the indicator information on planning decisions. Other benefits are that it does not contribute to the misconception that the well-being of society and the environment are independent, and usually in opposition, and that it places greater emphasis on the human behavioural changes that are fundamental to achieving a state of sustainability.

Future development of this indicator assessment tool will be focused on the development of the fuzzy rule-based system for integration of indicator information in the assessment. When development is complete, the tool will be capable of assisting urban planners with sustainability assessment of stormwater management scenarios.

References

ANZECC State of the Environment Reporting Task Force. (2000) Core environmental indicators for reporting on the state of the environment. Australian and New Zealand Environment and Conservation Council, Environment Australia.

Ashley, R., Souter, N., Butler, D., Davies, J., Dunkerley, J. and Hendry, S. (1999) Assessment of the sustainability of alternatives for the disposal of domestic sanitary waste. *Water Science and Technology* **39** (5), 251-258.

Auckland Regional Council. (2003) Stormwater management devices: Design guidelines manual. (TP10) Auckland Regional Council.

Australian State of the Environment Committee. (2001) Australia State of the Envrironment 2001. Independent Report to the Commonwealth Minister for the Environment and Heritage.

Balkema, A., Preisig, H., Otterpohl, R. and Lambert, F. (2002) Indicators for the sustainability assessment of wastewater treatment systems. *Urban Water* **4**, 153-161.

Bockstaller, C. and Girardin, P. (2003) How to validate environmental indicators. *Agricultural Systems* **76** (2), 639-653.

Bossel, H. (1996) Deriving indicators of sustainable development. *Environmental Modeling and Assessment* **1**, 193-218.

Butler, D. and Parkinson, J. (1997) Towards sustainable urban drainage. *Water Science and Technology* **35** (9), 53-63.

Connell Wagner. (2001) Flood Study and Stormwater Management Plan for Tooway Creek Catchment. Caloundra City Council.

A decision support tool for implementing a sustainable resource management in the sector of municipal wastewater treatment

T. Dockhorn and N. Dichtl

Institute for Sanitary and Environmental Engineering, Technical University of Braunschweig, Pockelsstrasse 2a, D-38106 Braunschweig, Germany (E-mail: t.dockhorn@tu-bs.de)

Abstract Undoubtedly, conventional systems for wastewater treatment have significantly improved living conditions since their introduction especially in urban areas and have distinctly improved environmental protection. Regarding the sustainability of conventional wastewater treatment technologies, it has to be considered that these end-of-pipe technologies are mostly incompatible with the requirements of a modern sustainable resource management. A main characteristic of the existing systems is an extensive resource demand in order to eliminate those resources contained in municipal sewage such as nutrients and energy. Therefore, based on the principle of materials flow management, concrete incentives should be established, which promote solutions connecting the positive aspects of the existing systems (reliable disposal) with the advantages of a sustainable resource management. In the sector of municipal wastewater treatment, especially the separation of highly concentrated partial streams (source separation) such as faeces and urine with a subsequent recycling represents a suitable approach. Beyond the aspect of resource recovery a stream separation concept in the sector of municipal wastewater treatment leads also to a significant unloading of the essential resource fresh water, especially in water limited regions. A comparison of conventional and source separating systems under resource economical aspects discloses an economic potential in favour of stream separating concepts which can be used as a control instrument for strengthening the implementation of sustainable solutions. An exemplary approach will be presented in a concrete case study for the treatment of municipal wastewater of 350,000 people equivalents (PE).

Keywords Materials-flow analysis, resource management, decision support tool, source separation, nutrient recovery

Introduction

According to the theory of cowboy economics by Kenneth Boulding the natural environment provides resources for the economic system and serves as a waste bin for residues. This concept originating from conventional industrial economy applies in principle also to the end-of-pipe treatment of municipal sewage. If, however, wastewater components are interpreted as misguided resources within a production process, they then must be considered as economically utilisable resources and the connection between economic and natural cycle can be closed according to the principle of spaceship economics (Kenneth Boulding).

Materials-flow management

The aim of a materials-flow management is the ecologically and economically efficient manipulation of materials-flow taking into account social aspects. The basis for materials-flow management is a comprehensive and intermedia analysis and an evaluation of all relevant materials-flows (Spengler, 1998). Based on the results of the materials-flow analysis, objectives have to be defined and necessary strategies to achieve these objectives have to be developed. Such an approach will contribute to a fundamental reduction of resource input along the entire economic process chain, to a reduction or substitution of ecologically harmful substances and to closing the natural resource cycle. This correlation points out that the development and application of a materials-flow management in the sector of municipal wastewater treatment inevitably includes the re-utilisation of the resources contained in wastewater.

Wastewater and resources

Wastewater per se is a resource collector since it contains most of the end-products of the human metabolism. Since human beings are last in the food chain, conventional wastewater treatment systems withdraw nutrient from the natural cycle thus interrupting the resource cycles.

Within the present case study the following resources will be considered:

- organic compounds (COD)
- nitrogen (N)
- phosphorus (P)
- potassium (K)

The study can principally be extended to other resources (e.g. S, Ca, Mg, Na, etc.) and potential pollutants such as heavy metals, halogenated organic compounds, endocrine substances or pathogens.

For a multicriteria assessment of the resource consumption of different wastewater treatment concepts, it is necessary to aggregate the various criteria (e.g. energy, resource equivalents, etc.) by using a common and comparable parameter (Zimmermann and Gutsche, 1991). Or else the problem might arise that one has to compare e.g. an option with intensive resource demand but extensive energy consumption to a second option which requires less resources but more energy. Such multicriteria assessments can be accomplished e.g. by using the outranking method PROMETHEE (Brans *et al.*, 1986). Since the present case study does not only consider the materials but also the energy flow, it is beneficial to aggregate relevant criteria prior to the assessment. One approach for instance determines for each process the equivalents of primary energy required and also calculates the consumption of resources (e.g. for production of phosphate fertiliser) in terms of primary energy equivalents (Maurer *et al.*, 2003). In the present case study all relevant criteria were aggregated using their monetary equivalent. The advantage of this approach is that a potential resource shortage will be reflected in the market price of the resource (price as a scarcity indicator causes adjustment) and will thus directly influence the evaluation. In contrast to that a shortage of resources such as phosphorus does not result directly in an increasing consumption of primary energy.

To determine the economic potential of resources contained in wastewater the theoretical market price of relevant resources will be evaluated as well as their value regarding the production of commercial complementary products.

Phosphorus

Considering the aspect of resource protection, phosphorus is the most important resource contained in wastewater, because the global phosphorus deposits are limited and even nowadays it can be foreseen that great efforts will have to be undertaken in future to ensure especially the supply of agriculture and industry with raw phosphorus by closing the open cycles. 140 Mt/a rock phosphate is mined worldwide and industrially processed to approx. 40 Mt/a of phosphate fertiliser and to 5 Mt/a of phosphate-based detergents (Gumbo *et al.*, 2002). Whereas the electro thermal process is intensive in energy consumption (the production of white phosphorus requires approx. 13 kWh/kg P), the chemical processing requires large amounts of sulphuric acid (Keudel *et al.*, 1999). During this case study, expenses for the industrial phosphorus processing are determined using the market price for commercially available phosphate fertiliser and are assumed at €0.782/kg P (DAINET, 1998). The phosphorus removal at wastewater treatment plants again requires resources. The consumption of primary energy equivalents for chemical phosphorus precipitation lies at 13.6 kWh/kg P (Maurer *et al.*, 2003). The used resources and the precipitated phosphorus are usually not recycled from the sewage sludge but disposed with it.

Nitrogen

The industrial processing of ammonia using the steam reforming process requires approx. 9.6 kWh/kg N of electrical energy and according to the stoichiometry of the process 570 g CH_4/kg NH_3-N. On the other hand the nitrification of nitrogen in a wastewater treatment plant theoretically requires 3.07 kWh electricity/kg N. Thus the utilisation of nitrogen from partial wastewater streams saves a total of 12.7 kWh electricity/kg N. Maurer *et al.* (2003) report a total energy demand for nitrification and pre-denitrification of about 12.5 kWh/kg N. In this case, the resource potential of nitrogen is calculated by the market price for nitrogen fertiliser and assumed at €0.706/kg N (DAINET, 1998).

Potassium

Potassium is an ubiquitary element, insofar as its impact in terms of resource protection is of secondary priority. Nevertheless it is an essential nutrient for crops and its resource potential is assumed by the market price for potassium fertiliser at €0.297/kg K (DAINET, 1998). Since the well soluble potassium accumulates in sewage sludge only to a minor amount the main proportion remains dissolved and thus the liquid phase would have to be used as fertiliser.

Organic compounds

Organic compounds contained in wastewater can be used energetically in an anaerobic treatment. Regarding the economic efficiency of this process, it is necessary to realise high initial substrate concentrations in the partial streams, since for example wastewater must be heated for the treatment. While the aerobic degradation of organic compounds, used at conventional wastewater treatment plants, requires approx. 0.35 kWh electricity/kg COD_{elim}, the anaerobic treatment with biogas production and a subsequent electricity generation produces approx. the same amount of energy. Thus the overall potential of an anaerobic instead of an aerobic treatment process amounts to 0.7 kWh electricity/kg COD_{elim} or even 2 kWh/kg COD_{elim} in terms of primary energy produced. During the present case study the following figures were used for the anaerobic and anaerobic treatment.

Table 1 Figures for the aerobic and anaerobic treatment of organic compounds

Aerobic treatment	ratio COD/BOD_5	2.3
	specific O_2-demand	1.2 kg O_2/kg BOD_5
	specific O_2-yield	1.5 kg O_2/kWh
	electricity supply conventional	€0.065/kWh
Anaerobic treatment	COD degradation	80 %
	specific methane production	0.15 kg CH_4/kg COD_{elim}
	energy content CH_4	10 kWh/ m^3
	electricity feed regenerative	€0.087/kWh
	energy efficiency gas engine	32 %

For the economic evaluation of the resources contained in wastewater, the figures given in Table 2 were used. Whereas the economic potential of the resources P, N and K was calculated by market prices of commercially available fertiliser products, for the determination of the resource potential of the COD an anaerobic treatment with a subsequent electricity generation was assumed according to figures given in Table 1.

Table 2 Economic equivalents used for the relevant resources

Parameter	Monetary resource equivalent
COD	€0.047/kg COD
P	€0.782/kg P
N	€0.706/kg N
K	€0.297/kg K

Materials-flow analysis

The basis for conducting a materials-flow analysis is the qualitative and quantitative description of the resources contained in municipal sewage. For the present case study the following partial wastewater streams related to 350,000 PE were analysed:

1. municipal wastewater (total stream)
2. separate yellow water (urine) collection
3. separate brown water (faeces) collection
4. grey water

The case study is based on the actual operation data and costs of a full-scale biological wastewater treatment plant for 350,000 PE. The partial stream of urine was assumed at 500 l urine/PE*a (waterless urinal) and the brown water stream at 50 l faeces /PE*a and 3 l/d fresh water for flushing (vacuum toilet). The percentage distribution of the different resources to the individual partial streams was assumed according to Otterpohl (2000). The following Table 3 shows the relevant resource concentrations and loads in the individual wastewater streams.

Table 3 Concentrations and loads of the different partial wastewater streams for 350,000 PE

Parameter	Load [t/a]	Proportionate load [%]	Concentration [mg/l]	Partial stream
COD	15,695	100	689	
TKN	1,409	100	61.9	
P_{tot}	190	100	8.3	total wastewater
SS	5,694	100	250	
K	630	100	27.7	
Q [m^3/a]	22,776,000	100		
COD	1,883	12	10,762	
TKN	1,226	87	7,005	yellow water
P_{tot}	95	50	543	(urine)
SS	0	0	0	
K	340	54	1,944	
Q [m^3/a]	175,000	0.8		
COD	7,377	47	18,407	
TKN	141	10	352	brown water
P_{tot}	76	40	190	(faeces)
SS	2,847	50	7,104	
K	76	12	189	
Q [m^3/a]	400,750	1.6		
COD	6,435	41	290	
TKN	42.3	3	1.9	
P_{tot}	19.0	10	0.9	grey water
SS	2,847	50	128	
K	214	34	9.6	
Q [m^3/a]	22,200,250	97.4		

Required efforts for conventional wastewater treatment

The collection of charges for the treatment of municipal wastewater is mostly based on the principle of commensurability. Annual costs for wastewater treatment (invest and operating costs) are usually related to the annual quantity of wastewater treated (charging according to the consumption of fresh water). Furthermore heavy pollution fees can be added for discharges which exceed the average concentrations of pollutants in the wastewater according to the increased usage of wastewater treatment facilities. The calculation of heavy pollution charges usually consists of a pollution independent proportion (e.g. personnel costs, utilisation of the sewer system, primary treatment, etc.) and pollution dependant proportion (e.g. energy for aeration, sludge disposal, volume of the biological reactor, etc.). A further breakdown of the different costs centres can be conducted by relating the costs of individual pollution dependant cost centres to the individual wastewater parameters COD, N and P. In principle, the entire cost centre accounting of a wastewater treatment plant is to be divided according to this procedure.

The calculation of heavy pollution charges can be used for a direct economic assessment of pollutant loads which exceed the usual average wastewater composition.

Within the scope of the present case study, the above-described approach was used to quantify the additional effort necessary for the treatment of different partial streams. The current approach implies that the overall effort for wastewater treatment can be reduced by exactly that monetary amount, if certain highly concentrated partial streams (e.g. urine) are separated from the total wastewater stream and recycled afterwards. To create a concrete incentive for the stakeholders, the system of municipal wastewater charging would have to be modified in such a way that a so-to-say "resource bonus" (in an equal amount as the high pollution charge determined) for separated and subsequently recycled partial streams is deducted from the general wastewater treatment charges.

Such a system only works, if an actual source separation with a subsequent recycling takes place. If only water saving technologies are realised, the collectivity of users will not benefit, because the effort for wastewater treatment in principle remains the same and specific treatment costs will rise.

The present case study divides the actual total costs for sewer system and wastewater treatment plant at the investigated location into pollution dependent and pollution independent costs for the parameters COD, N and P according to each individual costs centres.

At first pollution dependent surcharges z were calculated for relevant parameters according to Eq. (1) (ATV, 1990).

$$z_C = \frac{c_{hp} - c_{lc}}{c_{av}} \cdot \frac{costs_{wwtp}}{total\ costs} \cdot \frac{f_x}{100} \qquad (1)$$

with:

z_C	surcharge for the relevant parameter [€]
c_{hp}	concentration of the parameter in highly polluted wastewater [mg/l]
c_{lc}	limiting value of the parameter for charging a heavy pollution charge [mg/l]
c_{av}	average concentration of the parameter in municipal wastewater [mg/l]
$costs_{wwtp}$	annual costs of the wastewater treatment plant [€]
total costs	sum of the annual costs (WWTP, sewer system, etc.) [€]
f_x	pollution dependent portion of costs for the parameter [-]

The heavy pollution charge for the parameters COD, N and P can be calculated according to Eq. (2).

$$CH = ch + [(ch \cdot z_{COD}) + (ch \cdot z_N) + (ch \cdot z_P)] \qquad (2)$$

with:

CH	total added wastewater treatment charge [€/m^3]
ch	general wastewater treatment charge [€/m^3]
z_{COD}, z_N, z_P	pollution dependent surcharges for the parameters COD, N and P

Considering the actual costs for the wastewater treatment plant and their itemisation to the individual costs centres, the following calculation (Eq. (3)) is derived in the present case study. The actual specific treatment costs (including sewer system, etc.) amount to €1.15/m^3.

$$CH = 1{,}15 + \left[1{,}15 \cdot \frac{c_{COD} - 900}{690} \cdot 0{,}324 + 1{,}15 \cdot \frac{c_N - 80}{62} \cdot 0{,}081 + 1153 \cdot \frac{c_P - 12}{8{,}5} \cdot 0{,}027\right] \qquad (3)$$

Resource potential of the partial streams

In the present case study, the economic resource potential of the different partial streams was calculated separately for each resource on the basis of the figures given in Tables 2 and 3. The costs for the conventional treatment process were calculated additionally (Eq. (3)). All results are summarised in Table 4.

Table 4 Economic resource potential and conventional wastewater treatment for different partial streams

	Parameter	Municipal wastewater	Yellow water	Brown water	Grey water
resource potential	N [€/m^3]	0.04	4.95	0.25	0.00
	P [€/m^3]	0.01	0.42	0.15	0.00
	K [€/m^3]	0.01	0.58	0.06	0.00
	COD [€/m^3]	0.03	0.50	0.86	0.01
sum resource potential [€/m^3]		0.09	6.45	1.32	0.01
costs for wastewater treatment	N [€/m^3]	0	10.39	0.41	0
	P [€/m^3]	0	1.94	0.65	0
	COD [€/m^3]	0	5.32	9.44	0
	general costs [€/m^3]	1.15	1.15	1.15	0.82
sum treatment costs [€/m^3]		1.15	18.80	11.65	0.82
wastewater quantity 350.000 PE [m^3/a]		22,776,000	175,000	400,750	22,200,250
treatment costs 350.000 PE [€/a]		26,192,400	3,290,000	4,668,738	18,233,662

As shown in Table 4 the yellow water clearly dominates the actual resource potential and also the specific treatment costs. Although this partial stream contributes only 0.77% to the total quantity of wastewater, the proportion of its treatment costs amounts to 12.6%. Therewith yellow water, which is the main cost driver within this scenario, would be under resource economical considerations the preferential stream for a source separation with a subsequent recycling. Although phosphorus has the major relevance under the aspect of resource protection, nitrogen represents the highest economic resource potential at actual conditions of market economy. In the case of brown water, which represents the second most important partial stream, a volumetric proportion of 1.8% is related to proportionate total treatment costs of 17.8%.

Provided that the entire separation of yellow and brown water will lead to a decrease in the overall treatment costs (equal to the monetary efforts for the treatment of yellow and brown water shown in Table 4), the calculated costs for the treatment of grey water would decrease from €1.15/m^3 to €0.82/m^3. Thus, even if existing treatment facilities are used continuously, the reduced pollution load will open an economic margin which can be used for successively implementing stream separation processes.

In such cases, where a partial stream separation leads to a changed wastewater composition (see Table 3) and where it is possible to realise a new and adapted technology for the treatment of grey water (Wilsenach and van Loosdrecht, 2003) the monetary treatment efforts could be further decreased.

A re-dimensioning of the necessary activated sludge reactor volume (ATV-DVWK, 2000) using the annual values given in Table 3 in combination with a primary sedimentation (assumed at 0.5–1 h hydraulic retention time) leads to the relevant results shown in Table 5.

The comparison of the total investment costs for the different scenarios is based on the real investment costs for the actual treatment facilities (scenario 1). The investment costs for scenario 2 (mixture of grey and brown water) and scenario 3 (grey water) were calculated proportional to the calculated reactor volumes and the corresponding population equivalents. A modified calculation approach related to population equivalents (ATV, 1995) was used to estimate the investment costs for wastewater treatment for scenario 2 and scenario 3.

Table 5 Compilation of selected results from the re-dimension for three different scenarios

	Scenario 1 total wastewater	**Scenario 2 grey + brown water**	**Scenario 3 grey water**
treatment objective	COD elimination, nitrification, simultaneous denitrification, P-precipitation	COD elimination	COD elimination
reactor volume V_R [m^3]	49,958*	13,669	6,567
V_D /V_R [-]	0.45	0	0
precipitation product demand [kg Fe^{3+}/d]	533	0	0
ES_P [kg SS/d]	10,435	10,936	5,254
SRT_{tot} [d]	19	5	5
average O_2 consumption [kg/d]	31,012	15,318	6,186
NO_3-N effluent [mg/l]	12.6	0	0
P_{tot} effluent [mg/l]	2.0	1.7	0
comments	/	minimum factor N	minimum factor N, P
PE [-]	358,280	98,169	46,935
specific total investment [€/PE]	175	278	336
total investment WWTP [€]	62,700,000**	27,300,000	15,800,000

(*Existing reactor volume: 51.000 m^3; **real total investment; ES_P = excess sludge production; SRT_{tot} = total sludge retention time; V_D = volume of denitrification reactor.)

Balance sheet classification and distribution of costs centres

The chosen balance sheet classification and the distribution of costs centres will determine whether a stream separating concept with a subsequent recycling will lead to an economic incentive for the stakeholders or not. In principle, the balance sheet classification should comprise beside the wastewater treatment also the components drinking water supply and recycling of resources instead of disposal. Only under this constellation, benefits from theoretically existing synergies can be used. The following relevant cost centres can be identified within the scope of a materials-flow management in the sector of municipal wastewater treatment.

Table 6 Relevant cost centres within a municipal materials-flow management

Cost centre	Stakeholder	Economic potential
1. wastewater treatment	operating company; end user in households	reduction of investment and operating costs
2. resources in the partial streams	end user in households recycling company	commercialisation of resources
3. drinking water	end user in households	saving of drinking water

Within the chosen balance sheet classification all costs centres and stakeholders have to be connected with each other. Considering the existing classification, neither does the collectivity of end users benefit from water saving technologies nor does the user who utilises ecological sanitation techniques with a subsequent recycling benefit from reduced wastewater treatment costs provoked by him. The "resource bonus" introduced before which has to be allocated to the general wastewater treatment charges could be a promising control instrument for the enhanced implementation of stream separating concepts.

The following exemplary calculation can be conducted for a household of four (Table 7) for a separate urine collection with a subsequent recycling.

Table 7 Exemplary wastewater treatment charge calculation for a household of four

1. drinking water supply: 126 l/PE*d * 4 pers. * 365 d = 184 m^3/a; á 1.60 €/m^3	294.00 €
2. wastewater treatment charge: 184 m^3 /a* 1.15 €/m^3	+211.60 €
3. resources in urine: 4 pers * 0.5 m^3 /a * 6.45 €/m^3	-12.90 €
4. resource bonus: 2 m^3 * 18.80 €/m^3	-37.60 €
5. saving of drinking water: 4 pers. * 2 l/flush * 5 flush/d * 365 d = 14.6 m^3/a; á 1.60 €/m^3	-23.36 €
total	431.74 €/a

In contrast to the conventional system where the total costs for drinking water supply and wastewater treatment would amount to €505.60/a, the alternative scenario with a urine separation and the allocation of a resource bonus would lead to a calculative annual saving of 18.47 € per person. Calculated for 350,000 PE the overall economical potential would theoretically amount to 6.5 Mio. €/a.

Of course it can not be expected that this newly introduced approach will effect an ad hoc replacement of the existing systems for wastewater drainage and treatment. But the approach identifies an economic potential for successively introducing sustainable concepts in spite of the long term character of the existing systems. Especially locations where new sanitary and wastewater treatment facilities are realised or where stream separating solutions with a subsequent disposal of the resources already exist, are suitable for this purpose.

Conclusions

Beside its many attainments, conventional modern wastewater treatment represents an elimination of essential and partly non-renewable resources under resource economical aspects. A main criteria of the existing systems is that especially partial streams that contain high concentrations of utilisable resources are diluted to such an extend that a recovery of the resources is economically not reasonable anymore.

Ecological sanitation concepts that include a subsequent recycling may therefore outclass conventional systems.

Especially the long-term character of the existing systems with high investment costs and long amortisation periods makes their ad hoc replacement almost impossible.

The presented new approach to evaluate the economical potential of resources contained in the different partial wastewater streams in connection with the identification of the specific costs for conventional wastewater treatment of these streams, points out an economical margin, which can be used for a successive introduction of stream separating concepts with a subsequent recycling. Urine source separation and recycling has the major priority within this approach due to its high resource potential, the large necessary effort for its conventional wastewater treatment and the comparatively low quantity of this partial stream. Although phosphorus has a major relevance under the aspect of resource protection, nitrogen represents the highest economic resource potential at actual conditions of market economy.

To enhance the usability of the already existing economic potential, it will be a fundamental task to modify and to extend the actual systems of wastewater charging in such a way, that a pollution dependant charge is introduced and a refund using a resource bonus is made possible. This will create an economical incentive for individual stakeholders for an enhanced realisation of stream separating systems in combination with a reuse of the resources.

References

ATV (1990) Starverschmutzerzuschläge –Technisch-wissenschaftliche Grundlagen der Gebührenermittlung für industrielle Nutzer öffentlicher Abwasseranlagen. Arbeitsbericht des ATV-Fachausschuss 7.4; *Korrespondenz Abwasser* **9**, 1075–1079.

ATV (1995) Handbuch Betriebstechnik, Kosten und Rechtsgrundlagen der Abwasserreinigung, *Ernst&Sohn*, 4. Auflage, p. 486.

ATV-DVWK (2000) Bemessung von einstufigen Belebungsanlagen. Arbeitsblatt A131, Deutsche Vereinigung für Wasserwirtschaft, Abwasser und Abfall e.V., Hennef.

BRANS, J.P., Vincke, P. and Mareschal, B. (1986) How to select and how to rank projects; the PROMETHEE method. *European Journal of Operational Research* **24**, 228–238.

DAINET (1998) Deutsches Agrar-Informationsnetz. http://www.dainet.de/fnr/de/Leitfaden/kap4/4_11_8.pdf

Gumbo, B., Savenije, H.H.G. and Kelderman, P. (2002): Ecologising societal metabolism: the case of phosphorus. In: Proc. 3rd Int. Conf. Environmental Management, 27–30.08.2002.

Keudel, L., Müller, J. and Dichtl, N. (1999) Technische Lösungen zur Wert- und Schadstofffraktionierung von Klärschlämmen. Studie im Auftrag des Niedersächsischen Umweltministeriums, unpublished.

Maurer, M., Schwegler, P. and Larsen, T.A. (2003) Nutrients in urine: energetical aspects of removal and recovery. *Wat. Sci. Tech.* **48** (1), 37–46.

Otterpohl, R. (2000) Design of highly efficient source control sanitation and practical experiences. EURO-Summerschool DESAR, June 2000, Wageningen, The Netherlands.

Spengler, T. (1998) Industrielles Stoffstrommanagement: betriebswirtschaftliche Planung und Steuerung von Stoff- und Energieströmen in Produktionsunternehmen. Erich Schmidt Verlag, Berlin.

Wilsenach, J. and van Loosdrecht, M. (2003) Impact of separate urine collection on wastewater treatment systems. *Wat. Sci. Tech.* **48** (1), 103–110.

Zimmermann, H.J. and Gutsche, L. (1991) Multi-Criteria Analyse, Springer Verlag, Berlin.

Sustainability management by water utilities: views across Australia, Europe and the USA

V. Uhlmann

Advanced Wastewater Management Centre, University of Queensland, Australia

Abstract The aim of this research was to develop an affordable, practical and user-friendly toolkit for use by water utilities in defining, assessing and improving their sustainability. SUST was developed from a range of best practice concepts in the literature, and designed to balance the social, environment and economic aspects of the "triple bottom line" to improve sustainability.

Five SUST modules were developed to provide a systematic approach to decision-making within which sustainability could be defined, indicators chosen to measure progress and options for improvement chosen in a systematic and rigorous manner.

A water industry survey was conducted to identify current sustainability reporting and the usefulness of SUST in overcoming barriers to improvement. Subsequent case study trials further refined the tool and identified extent of regulator support.

Australian findings indicated little linkage between current utility performance reporting and sustainability concepts. Both utilities and regulators considered SUST a better approach to sustainability management than existing methods as it provides a clearer picture of sustainability, with the model definition the most useful part of the tool. Both parties also agreed that sustainability needs to be defined prior to choosing indicators. This aids in redefining utility core business to focus on sustainability. However, given the current state of our knowledge, some major cornerstones of sustainability – total water cycle management, public involvement & real value pricing - remain to be addressed.

Internationally, SUST is viewed as a systematic approach to sustainability management, useful at the strategic as well as the project level and also as a cultural change tool. It has the potential to be applied in different cultural settings and to generate a standardized set of sustainability indicators for water utilities.

Keywords Sustainability; water utilities; definition; indicators; decision steps

Introduction

What does sustainability mean for a water utility? And, assuming it can be defined, how does a utility measure and improve their performance in relation to this intangible concept?

This doctoral research was conducted at the Advanced Wastewater Management Centre, University of Queensland in Australia, with the aim of developing an affordable, practical and user-friendly toolkit for use by water utilities in defining, assessing and improving their sustainability. Multi-disciplinary in nature, SUST was developed from a range of best practice concepts in the literature, and designed to balance the social, environment and economic aspects typically for the "triple bottom line" concept used to improve sustainability.

Development of SUST

Seven principles of sustainability were proposed for a water utility as a basis for the development of SUST:

1. Balancing the triple bottom line of social, environmental and economic factors (Elkington 1999).
2. Total water cycle management (TWCM) (Loucks 2000; Marsalek and Savic 2001).
3. Participative decision-making with stakeholders (Gleick 1998; Hardi and Zdan 1997).
4. Least cost planning (White and Howe 1998).
5. Identification and incorporation of externalities in pricing (Bowers and Young 1999).
6. Multi-disciplinary approach (Loucks and Gladwell 1999).
7. Adaptive management (Holling 1978).

Externalities are the unintended and often intangible impacts, both positive and negative, on third parties and the environment (Bowers and Young 1999).

Based on these principles, five SUST modules were developed as an intended complete sustainability management package:

1. Sustainability management steps for systematic and rigorous decision-making;
2. Model definition of a sustainable water utility as a starting point for a sustainability vision;
3. Selection criteria for choosing sustainability indicators;
4. A model sustainability indicator set which could be used to measure progress against the definition;
5. Multi-criteria analysis (MCA) software for systematically and rigorously assessing sustainability and options for improvement.

The purpose of these modules is to provide an understandable step-by-step approach to sustainability management for a water utility, beginning with a model definition for transformation into a common sustainability vision for a utility, followed by a systematic method for choosing indicators and a rigorous and transparent method for assessing options for improvement, together with stakeholders.

Research questions

In order to test the hypothesis that SUST may be able to provide the range of tools required for rigorous water utility sustainability assessment and management, three research questions were developed and addressed:

1. What is the current status of sustainability reporting by water utilities in the Australian State of Queensland and what are the barriers to its improvement?
2. What is the usefulness of SUST in improving sustainability management?
3. Does the use of SUST by water utilities achieve regulator and public agency support?

Research method

Following development of SUST from best practice in the literature, a water industry survey was conducted with fourteen water utilities in the Australian State of Queensland to address Question 1 and provide initial data in relation to Question 2. Subsequent case study trials with two utilities, Citiwater Townsville and Gold Coast Water, along with their stakeholder regulators, and utilising a refined SUST, obtained more in-depth data on Question 2 and also addressed Question 3. Data collection was by means of surveys and interviews.

Australian findings

Consistent with other research (Gleick 2000; Jackson and Roberts 1997) , Question 1 findings unsurprisingly identified the major barrier to improved sustainability management as a lack of understanding of the concept, with current performance reporting bearing little relationship to cornerstone sustainability principles. Sustainability for a water utility was often described in terms of security of water supply, which is perhaps to be expected given the drought conditions which have increasingly affected Australia's water supplies. The broader triple bottom line concept of sustainability was not seen as part of core business for a utility, but more a "nice-to-have" if you have the resources. As a consequence, while the engineering skills in utilities provide for effective analysis of problems and solutions for improvement in performance, a sustainability vision or goal is often absent in the decision making framework.

As stakeholder participation is also a cornerstone of sustainability (Loucks and Gladwell 1999), best practice indicates that such a vision or goal needs to be developed participatively in order to educate, learn together and build relationships of trust, and ultimately to develop stakeholder ownership of solutions and behaviour change. However, while stakeholder participation was largely supported in principle by participating utilities, it was not without its detractors. Many thought it was difficult to achieve an effective outcome from stakeholder involvement, and some were decidedly uncomfortable with giving a voice to what was perceived as uninformed public opinion. Such resistance is understandable; however there are ways to overcome the difficulties. One approach to deal with conflicting stakeholder objectives is to provide information on sustainability principles with a view to obtaining agreement on those at the outset as a framework for subsequent more detailed and possibly difficult decisions (SUST Module 1: Steps 4 and 5). The multi-criteria analysis software in SUST Module 5 can also enable transparent trade-offs between conflicting objectives.

Initial feedback on Question 2 indicated that SUST may be able to assist in addressing the need for a sustainability decision framework. Survey respondents supported the foundation principles of SUST and most of its modules. Module 1 decision-making steps were largely supported, though concern was raised by some utilities about the meaningfulness of obtaining a single sustainability figure as part of the sustainability assessment. Some viewed it as useful only at the political level for communication with stakeholders, with a more detailed assessment being required for internal management purposes.

The model definition of a sustainable utility in Module 2 was also supported. It is believed that this is the first time such a detailed description of sustainability for a water utility has been developed, and it was the subject of much discussion by participating utilities. One of the case study trial utilities confirmed that SUST enabled the development of a common agreement on the definition of sustainability and subsequently helped redefined their core business to focus on sustainability.

Support was also achieved for the use of indicators as sustainability measures in Module 4, consistent with the view that indicators are a practical vehicle for coming to grips with sustainability (Farrell and Hart 1998; OECD 1993; Pannell and Glenn 2000). However, agreement on a standardized indicator set for common use by utilities was not reached. Only 2 of the 10 model indicators are commonly used in reporting by utilities. In this regard, we have a long way to go in sustainability reporting; and it is timely to remember that sustainability is a journey, not an end. With increasing knowledge and understanding, such indicators may be seen as more relevant and the data available for reporting.

While agreement was unable to be reached on a common set of sustainability indicators, the view was expressed by participants that SUST may be able to produce a standardized indicator set which both satisfies regulators as well as utility owners, avoiding duplication of reporting effort. Such an outcome could improve both the practicality and affordability of sustainability reporting by water utilities and merits further research.

More detailed feedback from the case studies established that most utility and agency participants found SUST very useful in making sustainability both meaningful and operational for a water utility, with the model definition ranked the most critical of the SUST modules. More than half the participating regulatory agencies thought SUST represented a better approach to managing sustainability than current methods, and would support its use by utilities, while the remainder was unsure.

Further, consistent with the literature (Hardi and Zdan 1997; Lundin 1999), participants believed that, to maintain validity, an indicator set needs to be developed following agreement on the definition of sustainability, and cannot be chosen without such a framework.

Three cornerstones of water utility sustainability remain as challenges: TWCM, public involvement and real value pricing. Institutional and political boundaries in Australia mean that many utilities face difficulties in achieving TWCM as they do not have control over the catchment or their bulk water supplies. Additionally, despite in-principle support for a multi-disciplinary approach, many utilities also do not possess in-house the social science skills required to manage regular public involvement in sustainability decisions. And finally, without the identification and costing of externalities involved in the sustainable provision of water supply and sanitation services, the real value of water cannot be calculated.

International feedback

During 2004, views were sought on the above research findings from a range of water utilities and research organisations in France, Germany, Portugal, Spain, Sweden, the United Kingdom and the United States.

International feedback obtained provided useful comment on both the concept of sustainability and SUST. Some respondents thought the term "sustainability" was over-used and abused and in need of being replaced, though only "flexibility and adaptability" were put forward instead. One discerning comment suggested the term should not be used if externalities, and therefore inter-generational equity, are not being addressed. The objective of achieving inter-generational equity is an important one in the Brundtland Commission's definition of sustainability (World Commission on Environment and Development 1987), so while SUST principles and strategies recognize the need to incorporate externalities to achieve real value pricing, a more explicit methodology is perhaps needed to address inter-generational equity.

There is a need to be more people friendly in developing sustainability approaches (Harris 2002). However, familiarity with the concept of balancing the triple bottom line varied a great deal across Europe and the USA, with many admitting social issues are not being well addressed.

The concept of total water cycle management, while recognized, is not well defined and this is hampering efforts in its achievement. A range of strategies seems to define TWCM, which is regularly interpreted in Australia as meaning effluent recycling, while stormwater collection and reuse may be more readily achieved in Europe.

The importance of stakeholder participation in decision-making was also recognized but not commonly undertaken beyond the basin management programs required by the EU Water Framework Directive.

While some individual SUST elements were acknowledged as not being new, the systematic approach of the overall package was seen as useful. It was viewed as being able to be applied at a number of levels - in project planning, strategic planning and in culture change. It could also be applied cross-culturally, though the participative component may need to be adapted. Some support was also obtained for SUST's role in helping develop a standard set of sustainability indicators for common use by water utilities and, in particular, drawing together the currently separate sets of IWA performance indicators for water supply and sanitation under a sustainability framework.

Major contributions of this research

The major contributions of this research are considered to be:

1. An improved method of sustainability assessment and reporting, SUST, which achieved support from the majority of participating regulatory agencies.
2. A complete step by step process whereby utilities and their stakeholders can agree on a vision of sustainability, then measure and make improvements.
3. A definition of a sustainable water utility which has been refined by participating utilities and agencies.
4. An educational tool for improving understanding of sustainability.
5. Recognition that, to be practical, affordable and user-friendly, a sustainability indicator set must also meet regulatory reporting requirements and, at least in the interim, traditional indicators may need to be incorporated, in contrast to Lundin's (1999) view that they should not be used.
6. Agreement that the water utility practice of selecting a set of sustainability indicators without first developing an agreed vision of sustainability is not effective.
7. Some evidence that a standardized sustainability indicator set could be developed using SUST and could be used by all utilities, which presents exciting prospects.
8. Evidence that many water utilities have a long way to go before they can address some of the cornerstones of water utility sustainability – total water cycle management, real value pricing and public involvement

References

Bowers, J. and Young, M. (1999) *Valuing externalities: a methodology for urban water*. Policy and Economic Research Unit, CSIRO Land and Water.

Elkington, J. (1999) Triple bottom-line reporting: looking for balance. *Australian CPA*, March, p. 3.

Farrell, A. and Hart, M. (1998) What does sustainability really mean? The search for useful indicators. *Environment*, **40** (9), 4–9, 26–31.

Gleick, P.H. (1998) Water in crisis: paths to sustainable water use. *Ecological Applications*, **8** (3), 571–579.

Gleick, P.H. (2000) The changing water paradigm: a look at twenty-first century water resources development. *Water International*, **25** (1), 127–138.

Hardi, P. and Zdan, T. (eds) (1997) *Assessing Sustainable Development: Principles in Practice*, 1 vols., International Institute for Sustainable Development (IISD), Winnipeg, Manitoba, Canada.

Harris, G. (2002) *Ensuring sustainability - paradigm shifts and big hairy goals*. CSIRO, Melbourne, April 8, 2002, Speech given at Enviro2002 Conference, Melbourne. www.csiro.au/index.asp?type=blankandid=Enviro2002_GrahamHarris.

Holling, C.S. (1978) *Adaptive Environmental Assessment and Management.* Wiley IIASA International Series on Applied Systems Analysis, John Wiley and Sons, Toronto.

Jackson, T. and Roberts, P. (1997) Greening the Fife economy: ecological modernization as a pathway for local economic development. *Journal of Environmental Planning and Management*, **40** (5), 615–629.

Loucks, D.P. (2000) Sustainable water resources management. *Water International*, **25** (1), 3–10.

Loucks, D.P. and Gladwell, J.S. (eds) (1999) *Sustainability criteria for water resource systems*. International Hydrology Series, Cambridge University Press, Cambridge.

Lundin, M. (1999) Assessment of the environmental sustainability of urban water systems. Licentiate thesis, Chalmers University of Technology.

Marsalek, Q.R. and Savic, D. (2001) Urban water as a part of integrated catchment management. In C. Maksimovic and J.A. Tejada-Guibert (eds), *Frontiers in urban water management: deadlock or hope*, pp. 37–83. IWA Publishing, London.

OECD (1993) *OECD Core Set of Indicators for Environmental Performance Reviews*, No. 83, OECD, Paris.

Pannell, D.J. and Glenn, N.A. (2000) A framework for the economic evaluation and selection of sustainability indicators in agriculture. *Ecological Economics*, **33**, 135–149.

White, S.B. and Howe, C. (1998) Water efficiency and reuse: a least cost planning approach. *Proceedings of the 6th NSW Recycled Water Seminar*.

World Commission on Environment and Development (1987). *Our common future*. Oxford University Press, Oxford, England, and New York.

GLOWA Jordan River Project: integrated research for sustainable water management

Holger Hoff*, Heike Küechmeister* and Katja Tielbörger**

*Department of Plant Ecology and Nature Conservation, University of Potsdam, Germany
**Department of Plant Ecology, University of Tübingen, Germany

Abstract The Jordan region is facing severe and increasing water scarcity. Solutions to the emerging water crisis are still very fragmented and subject to the political situation. The GLOWA Jordan River project provides scientific support for integrated water resources management. A key component of the project is the integration of water and land management, using the so-called "blue-green water concept" which integrates surface and groundwater fluxes with evapotranspiration from vegetation. Based on climate and other scenarios, that are developed jointly with stakeholders, different adaptation options are tested for their eco-hydrological impacts and costs and benefits, using the full range of scientific tools from experiments, monitoring, simulation models to socio-economic surveys and assessments. Eventually information from the GLOWA project is integrated with other regional knowledge in a central decision support tool.

Keywords Global change scenarios; water use conflicts; adaptive management

Introduction

Global change and globalization pose additional threats to save water supply in addition to increasing local pressures in many regions of the world. Common expressions of the looming water crisis are decreasing water availability, pollution and increasing demand, as well as higher climate variability and more extreme events (see e.g. WWDR 2003). Given the key role that water plays for humans and ecosystems, effects of water scarcity include hunger, disease, poverty, and ecosystem degradation. The complex interactions of local regional and global pressures on water resources and feedbacks of water scarcity on other sectors are best addressed in the framework of integrated water resources management (IWRM, see GWP 2000).

Sustainable water management requires a combination of complementary technical, economic, social and political measures, adapted to the respective local contexts (GWP 2000). The German Ministry for Education and Research (BMBF) is responding to this challenge with a concerted research effort in support of sustainable water management, the GLOWA programme. This programme has been implemented in a transect of five integrated regional projects in Europe (Elbe, Danube), Middle East (Jordan River) and Africa (Volta, Draa / Oueme). These projects develop, jointly with stakeholders, integrated strategies for water management under global change, adapted to the regional socio-economic and ecological conditions. Here, we present the approach and concept of the GLOWA Jordan River Project.

The water situation in the Jordan River region

Like in many semi-arid regions, high spatial and temporal climate variability pose severe challenges to water management, e.g. pronounced dry seasons, high inter-annual variability in precipitation and steep gradients of water availability away from the coast and from the highlands.

Water scarcity in the Middle East is among the highest in the world. Groundwater and surface water resources are fully committed and sometimes overexploited. For example, per-capita water availability in Israel is currently at 265 m^3 per year and in Jordan at 170 m^3 (WRI 2004), which is below all thresholds of water scarcity (e.g. 1700 m^3 or 1000 m^3 per capita and year, Falkenmark 1999).

Regional water scarcity is expected to increase further, due to internal and external pressure, in particular:

- decreasing water availability, due to climate change effects, such as increasing temperatures and higher climate variability (Alcamo 2003, Arnell 2004);
- increasing water demand, given that population growth rates are among the highest in the world (WRI 2004).

Conflicts over water are triggered also by unequal distribution of water resources in the region. This inequity is partially due to natural and climatological factors, but also to the political and economic situation. Consequently, water use conflicts often have a political component or political implications (Allan 2003, Jägerskog 2001). Sustainable solutions to the pressing water crisis are difficult to achieve under the current political situation. However, progress in resolving water issues can also facilitate the overall peace process. The special role of the water sector is also evident from the continuous activities of the Palestinian-Israeli Joint Water Committee and the agreement of both parties to keep water infrastructure intact, despite the conflict.

Adaptation to water scarcity and high climate variability has a long tradition in the region, which has been termed "cradle of agriculture" (Issar 2004). Water harvesting and irrigation have been applied for many centuries of even millennia, drought resistant crops have been used, and water productivity in irrigated agriculture has increased several times in recent decades (Tal 2004). The region is leading in technologies such as drip irrigation, green houses, wastewater reuse and more recently also seawater desalination.

For example, Israel strongly promotes the development of non-conventional water resources or „manufactured water", in particular the reuse of wastewater and desalination, to overcome water scarcity. By the end of this decade, reuse of wastewater is expected to exceed the use of freshwater in agriculture. In parallel a number of desalination plants will be operated along the Mediterranean coast with a total capacity of about 500 million m^3 per year (Water Commission 2003).

Jordan is rethinking a Red Sea – Dead Sea canal to increase water supply, in addition to demand management measures (El-Naser 2004). In this canal, the difference in elevation (sea level to 400 m below sea level) could be used to desalinate the water.

Despite all measures taken so far, the region "has run out of water" in the 1970s and can not be food self sufficient any more (Allan 2001). Adapting to this water scarcity, the region has more than any other region replaced local food production with food imports. The food demand of Israel and Jordan is mostly met through annual food imports, that require 2-3 times more water for their production than the total renewable resources of these two countries (Chapagain and Hoekstra 2004).

Despite the large number of strategies, plans and projects, supported by regional governments and also by development cooperation, the water crisis is far from being solved.

Therefore, within the integrated water resource management (IWRM) concept, water research needs to consolidate and improve the knowledge base, in particular to provide a better understanding of the complex interactions between natural processes and human resource use. Interdisciplinary and integrated projects are required to foster the application of scientific knowledge for maximizing the joint benefit of scarce water resources.

To this end, social, economic and environmental costs and benefits of different adaptation options need to be assessed simultaneously. Traditional disciplinary research approaches cannot provide the required comprehensive assessment of pressures, impacts and responses, because they focus on single aspects of water availability, demand or quality. The GLOWA Jordan River Project addresses this gap between science and application in water management.

The GLOWA Jordan River Project

The German Ministry for Education and Research (BMBF) has initiated the GLOWA (Global Change and the Hydrological Cycle www.glowa.org) programme, as an integrative science approach towards sustainable management of larger river basins (Rieland 2004).

In the youngest of the five GLOWA projects, GLOWA Jordan River (www.glowa-jordan-river.de), Israeli, Jordanian, Palestinian and German scientists jointly analyze the regional water crisis and possible solutions under current and global change conditions, in a set of complementary activities.

The first phase of this project (until mid 2005) has provided better process understanding and described the current situation, in particular the interactions and feedbacks of different aspects of regional water availability, demand and quality. Key trends of global change and globalization and their effects on regional water resources have been identified (e.g. Alpert 2004, Shaheen et al 2005). The system understanding gained in phase 1, provides a basis for the second phase of the project (2005-2008), in which scenario-based management options will be developed and assessed.

Close collaboration with different partners and stakeholders from water management, politics, development cooperation and NGOs helps to bridge towards application.

The GLOWA JR project has five work-packages (Figure 1). Through coupling of all work-packages and integration of results in a central decision support tool, all components contribute to the central goal, i.e. scientific support for sustainable management of regional water resources. The five work-packages are:

1. Global Change Scenarios
2. Water Resources, Availability, Demand and Quality
3. Water in Natural Ecosystems
4. Water in Agriculture
5. Integration and Application

In the following, we briefly outline the main approaches and results of these work packages.

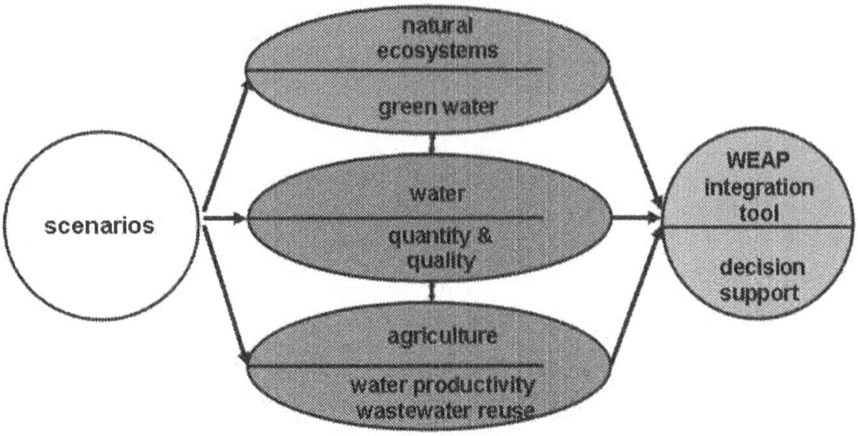

Figure 1 Integration scheme for work-packages 1–5 in the GLOWA Jordan River Project

Global change scenarios

All work-packages of GLOWA JR are scenario-based. The first phase has focused on the development of regional climate scenarios as a basis for all other activities. These high-resolution scenarios are driven by global climate models, employing statistical and dynamical downscaling methods (Kunstmann 2003, Boehm et al 2004). By coupling these scenarios with hydrological models, future water availabilities are calculated. These key drivers enter all other work-packages and projects.

Besides improved regional climate scenarios, phase 2 will also develop additional scenarios, in particular socio-economic scenarios. Initially, these will be qualitative scenarios or "storylines", developed iteratively between stakeholders and scientists. The stakeholders identify plausible future trends and key decisions in water management, and GLOWA science validates, refines and quantifies the initial storylines, employing experiments, monitoring and simulation models. Relevant scenarios in GLOWA JR may address land use, agricultural subsidies, investments and new technologies, but also international trade or life-styles. Relevant land use trends include changes in agricultural areas, including irrigated and grazing land, as well as protected areas and urban areas and their respective effects on hydrology and water resources.

Water resources, availability, demand and quality

In the first phase of GLOWA JR, water and constituent fluxes have been quantified for selected sub-catchments in the upper Jordan River and in the lower Jordan River, i.e. in side-wadis east and west of the lower Jordan River. Existing data on surface and groundwater resources have been compiled (e.g. Litaor et al 2004).

The second phase will build upon these achievements and results from other initiatives, by consolidating information in a central decision support (WEAP, Water Evaluation and Planning) tool and GIS data base.

When consolidating hydrological data, so-called "green" water fluxes (evapotranspiration from soils and plants will be integrated with "blue" (surface and ground) water fluxes (Falkenmark and Rockström 2004) for a full balance of all available water and the respective uses, including water fluxes in natural ecosystems, rainfed agriculture and grazing land. Green water provides the central link between water and land management. By taking into account green water fluxes, new scenarios of integrated land use and water use by humans and ecosystems will be developed.

Horizontal and vertical water fluxes are coupled and regionalized in GLOWA JR to the full basin scale. From that, spatial and temporal variability of water availability can be derived for the upper and lower Jordan River catchments for specified regional climate scenarios. Subsequently, the ecological and socio-economic implications of different management options for natural ecosystems and rainfed and irrigated agriculture will be assessed.

Water in natural ecosystems

The GLOWA JR project area comprises four biogeographic regions (Shmida and Aaronson 1986). The climatological boundaries of these biogeographic regions are largely determined by availability and temporal dynamics of precipitation. Hence natural ecosystems are likely very vulnerable to climate change. Vulnerability of natural ecosystems is a focus of GLOWA JR. GLOWA JR integrates bioclimatic models with mechanistic ecological and hydrological models, and analyses feedbacks between climate, land use changes, flora and fauna. Subsequently, socio-economic consequences of ecosystem changes are analyzed by assessing and changes in ecosystem services and by using e.g. willingness-to-pay approaches.

During the first phase of GLOWA JR, the effects of climate change (in particular changes of precipitation) on structure and composition of different ecosystems have been analyzed along a steep regional precipitation gradient (e.g., Holzapfel et al. in press, Petrů et al. in press). Manipulative experiments, in particular rain-exclusion and artificial rain experiments, are complemented by spatially explicit vegetation models (Jeltsch and Moloney 2002), follow a space-for-time approach, which uses the spatial gradient as a substitute for long-term climate changes (Dunne et a.l 2002).

From the water resources perspective, natural ecosystems compete with rainfed agriculture or grazing for scarce (green) water. Hydrological effects of different land management options (protection, grazing, afforestation etc.) will be regionalized in phase 2, e.g. with respect to water consumption, runoff generation or groundwater recharge.

Integration with the agricultural work-package of GLOWA JR is achieved through coupled ecological and hydrological experiments and modeling, as well as socio-economic and ecological cost-benefit assessments of different management options. The goal is to maximize the overall benefit derived from the scarce green and blue water resources.

Water in agriculture

Agriculture consumes the majority of all water in the Jordan basin, i.e. about two thirds of all (blue) water resources (GTZ 1998) and even larger volumes of green water. At the same time, agriculture and irrigation are the cause of considerable loads of salts, nutrients and pesticides in surface and groundwater. In phase 1 of GLOWA JR, nutrient transport in agricultural drained land have been analyzed and the fate of organic substances from reused wastewater and of steroid hormones has been monitored (e.g. Shore et al 2004).

The second phase will assess different agricultural management options under a range of global change scenarios, in order to minimize adverse impacts for humans and nature, and to maximize the overall benefit of scarce water resources.

One focus will be on reuse of wastewater for agricultural irrigation, integrating treatment level, resulting water quality, vulnerability of receiving soils and groundwater to contamination, and costs and benefits for farmers, consumers and environment.

Another focus will be on green and blue water productivity in irrigated and rainfed agriculture.

Given the expected reallocations of blue water from agriculture to urban and other water users (Taha et al 2004, Tal 2004), a combination of adaptation options will be required, in order to ensure food security. This will include on one hand increases of water availability, use of non-conventional water (Tal 2004), i.e. reused wastewater, desalinated water and harvested rainwater, but also increases in water productivity, i.e. biomass production per liter of water used (Oweis 2003), and further substitution of local food production through virtual water imports, i.e. imports of food (Allan 2003).

Integration

By integrating results from the above work-packages, GLOWA JR will improve the understanding of global change effects and tradeoffs between different adaptation options, when trying to maximize benefits for humans and nature through integrated water resource management.

Scientific integration and bridging to water management is facilitated by

(a) joint development of scenarios with scientists and stakeholders (SAS approach, Alcamo 2001), and

(b) cost-benefit analyses of various water and land management options (e.g. optimization of water allocations, reuse of wastewater, land use planning, substitution of agricultural production with food imports, and

(c) synthesis of results in a geographically explicit decision support tool (WEAP). This allows water managers and other users to simulate the hydrological and economic consequences of different global change and management scenarios (Yates et al 2005).

Conclusions

Given the pressing and ever increasing water scarcity in the region, a portfolio of adaptation measures is required to ensure water security for humans and sustainability of ecosystems.

Besides the standard measures of increasing supply – e.g. through desalination and wastewater reuse - and managing water demand in agriculture and municipal systems, major potential for improvement lies in the integration of blue and green water use.

The GLOWA JR project supports this concept through well coordinated experiments, model simulations and cost-benefit assessments, in a collaboration of Jordanian, Palestinian, Israeli and German scientists from a wide range of socio-economic and biophysical disciplines.

All of these GLOWA activities are based on a set of realistic, quantitative scenarios, which are developed jointly between GLOWA scientists and stakeholders. The resulting scenario-based information will be synthesized for application in water management in a central decision support tool (WEAP). With that, GLOWA JR helps to operationalize the principles of Integrated Water Resources Management in the region, and supports cooperation over shared water resources.

References

Alcamo, J. (2001) Scenarios as tools for international environmental assessments, European Environment Agency, Environmental issue report No. 24.

Alcamo, J., Maerker, M., Floerke, M. and Vassolo, S. (2003) Water and climate: a global perspective. Kassel World Water Series No. 6, Center for Environmental System Research, University of Kassel, Germany.

Alpert, P. (2004) The water crisis in the E. Mediterranean and relation to global warming? In: *Water in the Middle-East and N. Africa* (eds. F. Zereini and W. Jaeschke), pp. 55–61. Springer Verlag.

Allan, J.A. (2003) Virtual water eliminates water wars? A Case Study from the Middle East, Value of Water Research Report Series No.11, IHE, Delft, The Netherlands.

Allan, J.A. (2001) *The Middle East Water Question*, Tauris Publishers, London, UK.

Arnell, N.W. (2004) Climate change and global water resources: SRES emissions and. socio-economic scenarios. *Global Environmental Change* **14**, 31–52.

Boehm, U., Kuecken, M., Hauffe, D., Gerstengarbe, F.-W., Werner, P.C., Flechsig, M., Keuler, K., Block, A., Ahrens, W. and Nocke, T. (2004) Reliability of regional climate model simulations of extremes and of long-term climate. *Natural Hazards and Earth System Science* **4**, 417–431.

Chapagain, A.K. and Hoekstra, A.Y. (2004) Water footprints of nations. Value of Water, Research Report Series No. 16, IHE, Delft, The Netherlands.

Dunne, J.A., Saleska, S.R., Fischer, M.L. and Harte, J. (2004) Integrating experimental and gradient methods in ecological climate change research. *Ecology* **85**, 904–916.

El-Naser, H. (2004) Presentation at the International Water Demand Management Conference, Dead Sea, Jordan, June 2004.

Falkenmark, M. (1999) Forward to the future: a conceptual framework for water dependence. *Ambio* **28** (4), 356–361.

Falkenmark, M. and Rockström, J. (2004) *Balancing Water for Humans and Nature*. Earthscan, London, UK.

GWP (2000) Integrated Water Resources Management, Global Water Partnership, TAC Background Paper No. 4, www.gwpforum.org/gwp/library/TACNO4.PDF .

GTZ (1998) Middle East Regional Study on Water Supply and Demand Development, Evaluation Report, Eschborn, Germany.

Holzapfel, C., Tielbörger, K., Parag, H.A., Kigel, J., and Sternberg, M. (in press) Annual plant-shrub interactions along an aridity gradient. *Basic and Applied Ecology*.

Issar, A. and Zohar, M. (2004) *Climate Change - Environment and Civilization in the Middle East*. Springer Verlag, Heidelberg.

Jägerskog, A. (2001) The Sanctioned Discourse – a Crucial Factor for Understanding Water Policy in the Jordan River Basin, SOAS, University of London, Occasional Paper No 41.

Jeltsch, F. and Moloney, K. (2002) Spatially-explicit vegetation models: what have we learned? In: *Progress in Botany* (Esser K., Luettge U., Beyschlag W. and Hellwig F. eds), **63**, pp. 326–343. Springer Verlag, Berlin.

Kunstmann, H. and Jung, G. (2003) Investigation of feedback mechanisms between soil moisture, land use and precipitation in West Africa. In *Water Resources Systems – Water Availability and Global Change*, (Franks S. *et al.* eds), IAHS Redbooks Publication.

Litaor, M.I., Reichmann, O., Auerswald, K. and Shenker, M. (2004) The geochemistry of phosphorus in an altered wetland. *Soil Sci. Soc. Am. J.* **68**, 2078–2085.

Oweis, T. (2003) Comprehensive assessment on water management in agriculture, IWMI Water Policy Briefing Issue 8.

Petrů, M., Tielbörger, K. , Belkin, R., Sternberg, M. and Jeltsch, F (in press) Life history variation of an annual plant under two opposing selective forces along a steep climatic gradient. *Ecography*.

Rieland, M. (2004) Das BMBF-Programm GLOWA: Instrumente fuer ein vorausschauendes Management grosser Flusseinzugsgebiete. *Hydrologie und Wasserbewirtschaftung*, **48** (2), 83–85.

Shaheen, H., Jayyousi, A., Shadeed, S. and Jarrar, A. (2005) Hydrograph estimation in using GIS supported GIUH model, West Bank as a Case Study. In: *Proceedings of the International Water Conference*, Sharm Al-Sheikh, Egypt, March 2005.

Shmida, A. and Aronson, J.A. (1986) Sudanian elements in the flora of Israel. *Annals of the Missouri Botanical Garden* **73**, 1–28.

Shore, L.S., Reichmann, O., Shemesh, M., Wenzel, A. and Litaor, M.I. (2004) Washout of accumulated testosterone in a watershed. *Science of the Total Environment* **332**,193–202.

Taha, S. Sharkawi, R. Haddadin, N. Naber, L. and Magiera, P. (2004) Projection of municipal, industrial, touristic and irrigation demands in Jordan. Presentation at the International Water Demand Management Conference, Dead Sea, Jordan, June (2004).

Tal, S. (2004) Water development and management in Israel and the region. Presentation at the Stockholm Water Week 2004.

Thomas, C.D., Cameron, A., Green, R.E., Bakkenes, M., Beaumont, L.J., Collingham, Y.C., Erasmus, B.F.N., de Siqueira, M.F., Grainger, A., Hannah, L., Hughes, L., Huntley, B., van Jaarsveld, A.S., Midgley, G.F., Miles, L., Ortega-Huerta, M.A., Peterson, A.T., Phillips, O.L. and Williams, S.E. (2004) Extinction risk from climate change. *Nature* **427**, 145–148.

Water Commission (2003) Transitional master plan for water sector development in the period 2002–2010.

WRI (2004) World Resources Institute, Earthtrends, www.earthtrends.wri.org.

WWDR (2003) Water for People, Water for Life, UN World Water Development Report, UNESCO.

Yates, D., Sieber, J., Purkey, D. and Huber-Lee, A. (2005) WEAP a demand, priority, and preference driven water planning model: Part 1, Model characteristics. *Water International* (submitted).

Centralised versus decentralised wastewater systems in an urban context: the sustainability dimension

Goen Ho and Martin Anda

Environmental Technology Centre, Murdoch University, Perth, Western Australia 6150, Australia

Abstract Decentralised wastewater systems in an urban context in the developed world have the same investment and operating and maintenance costs as centralised systems. Water and nutrient recycling or reuse are more conducive with decentralised systems and they are therefore environmentally more sustainable. The public health outcome is comparable with centralised systems. Acceptance of decentralised systems requires wider dissemination of their sustainability advantages, a change in government policy and a regulatory framework for their implementation and management. Capacity building within the industry and for developers, builders and their consultants is essential. The sustainability of decentralised systems should be placed within the context of sustainable urban water management of water supply, wastewater treatment and stormwater run-off, and of city planning. This will contribute to making cities more sustainable in other sectors as well (energy, transport, food, building, biodiversity, waste).

Keywords Centralised; decentralised; onsite; small scale; sustainability; urban; wastewater

Introduction

When we responded to the call for papers for the Leading Edge Conference on Sustainability in Water-Limited Environments 2004, our intention was to present a summary of the outcome from a panel discussion on the subject held at the end of the joint International Water Association 6th Specialist Group on Small Water and Wastewater Systems Conference and the 1st International Conference on Onsite Wastewater Treatment & Recycling on 12–14 February 2004 in Perth, Western Australia. Fortunately for us the outcome from the panel discussion has been reported by Peter Spencer of the Water Corporation of Western Australia in the August 2004 issue of Water (the journal of the Australian Water Association). In addition Chris Davis reported on the conference itself in the May 2004 issue of Water. Chris observed that over 230 enthusiastic delegates from 35 countries participated with 127 platform and 30 poster papers presented. There was vigorous debate at the end of conference panel discussion.

Our intention in this paper is therefore to reflect on the papers presented at the above conference, the discussion that took place at paper presentations as well as at the panel discussion and deduce a number of principles on the sustainable dimension of our urban wastewater systems. We would also like to share our experience in the planning, development and implementation of estates designed with a sustainability marketing edge in Perth, Western Australia.

Urban context

It is desirable, however, to recap the context of the debate on the future choice of urban wastewater systems, i.e. whether it should continue to be centralised or move to decentralisation, the criteria on which we will base our choice, and the outcome based on our knowledge and experience thus far, before presenting our view on what needs to be changed if we want to follow the more sustainable path.

The context set for the panel discussion was a city with a population of 500,000+, not surprisingly similar to that of Perth (1.2 million people served by 3 centralised wastewater collection and treatment systems with disposal of treated wastewater through ocean outfalls). Management of operation, maintenance and upgrades are also typically centralised. This context is, however, typical of a city in a developed world, and our generalisation should be able to be extended to similar cases.

As the population of the city grows peri-urban and rural communities are engulfed by urban encroachment, and inner urban areas are redeveloped. Options for these two development scenarios are to connect them to a centralised system, or to retain and upgrade the decentralised system existing in a peri-urban area, or creating an independent decentralised system in an inner city redevelopment. Placed in this context a decentralised system is assumed to serve <5000 population. This is two orders of magnitude smaller than for a centralised system, but still one order of magnitude greater than the definition for small systems arbitrarily set by the IWA Specialist Group for Small Water and Wastewater Systems as systems treating less than 100,000 L per day. Depending on per capita water use within the house this is equivalent to wastewater produced by about 500 people at 200 L/p/day. A decentralised system can of course consist of many individual onsite systems or a series of larger clusters or one decentralised system.

The urban context described above is the norm in the developed world, whereas in the developing world the opposite is generally the case. In the latter wastewater from houses, businesses, and industry are treated onsite, and discharged (whether treated or not) to the ground or nearby drains and water courses. The question facing the communities in the developing world is, however, the same, i.e. whether they should install a centralised or decentralised system if they want to deal with their wastewater.

Sustainability dimension

There is general agreement that to achieve sustainability (whatever fine distinctions there are in definitions that have been put forward), we need to consider economic, environmental and social factors.

Economics

Let us first consider the economics of centralised and decentralised wastewater systems. It has become clear that properly designed, constructed, operated and maintained centralised and decentralised systems to achieve the same performance standard (let us say 20 mg/L BOD and 30 mg/L SS) cost about the same. In the Australian urban context the investment cost is between Aus$ 5,000 and 10,000 per house connection, and the annual operation and maintenance costs are between Aus$ 200 and 800 per house connection. Table 1 shows a cost comparison between a large centralised system and the smallest decentralised system (onsite unit).

Table 1 Cost comparison between centralised and onsite systems

Centralised system	**Onsite system**
Investment cost[1] Aus$ 5,000 to 10,000 per property (up to 80 % in sewerage system – pipes and pumps). Economy of scale possible, but also dis-economy of scale with deep sewerage over long distances requiring pumping.	Investment cost[2] Aus$ 5,000 to 10,000 (mainly in the treatment unit and reuse or disposal land area). Economy of scale possible with clusters of units.
Operation & maintenance costs[3] Approx. Aus$ 200-800 per property per year (again mainly in operation and maintenance of the sewerage system).	Operation & maintenance costs[4] Approx. Aus$ 400-600 per property per year by a service provider (mainly operation and maintenance of treatment unit and reuse or disposal land area).

Notes: 1. National Lifestyle Villages (2005); 2. Biomax (2005); 3. Water Services Association of Australia (2000); 4. Biomax (2005).

It has also become clear that these costs are a small fraction of the health burden costs, if we did not have these systems in place with the resultant endemic presence and epidemic of water related diseases (UNEP, 2002, Hutton & Haller, 2004). Hutton and Haller (2004) estimated that the benefits to costs ratio of providing water and sanitation in developing countries ranges from 5 to 28 from the following benefits:

(1) Health sector benefit due to avoided illness
(2) Patient expenses avoided due to avoided illness
(3) Value of deaths avoided
(4) Value of time savings due to access to water and sanitation
(5) Value of productive days gained of those with avoided illness
(6) Value of days of school attendance gained of those with avoided illness
(7) Value of child days gained of those with avoided illness

There is economy of scale to be realised with centralised systems. This is particularly so with the cost of treatment. The investment, operation and maintenance costs of centralised systems are, however, largely tied to the sewerage system. There can therefore be dis-economy of scale if pumping is required to move wastewater through the sewerage system, and if ingress of stormwater or groundwater is excessive when we try to collect and transport wastewater from a large urban area (hence long distances).

Contrary to expectations arising from the perception of a simple septic tank system as an onsite unit, which is seldom serviced except when there is blockage due to accumulation of sludge in the tank, onsite systems when properly designed, constructed and operated to conform to Australian Standards (AS 1547:2000) are comparable in capital and operating costs as a connection to a centralised system. Regular maintenance is also required even for septic tanks and in particular for more developed onsite units. It is now recognised that mandatory maintenance is necessary for onsite and decentralised systems. Western Australia leads the way in this regard by requiring that 'aerobic treatment units' be serviced quarterly by the manufacturer or qualified service provider. This has resulted in the coinage of the slogan 'decentralised systems, but centralised management', though what is referred to as centralised management is not necessarily by a single provider. Based on servicing contracts for onsite systems in Western Australia the service cost per year is comparable to the sewerage rate for the centralised system.

Based on equal performance the investment and operation & maintenance costs of centralised and decentralised systems are therefore comparable. Decentralised wastewater treatment technology is sufficiently advanced not to be the limiting factor in implementing decentralised systems, although it can benefit from further research. If investment in research into small scale and decentralised technology reaches the investment that has been put into centralised technology in the last 50 years, then greater efficiency can be achieved because of the different ways in which small scale technology can be designed and operated compared to the almost single dominance of the deep sewerage and activated sludge type treatment used for centralised systems.

The costs for either system are not small, but the public health (noted above) and environmental costs of not putting them in place are much greater, and this underpins the economics of having a wastewater treatment system in a community.

Environment

Environmental considerations indicate that small (onsite and decentralised) systems are more conducive to achieving environmental sustainability. These include water reuse, recycling of nutrients (nitrogen and phosphorus) contained in the wastewater, separating domestic wastewater from stormwater run-off, and excluding industrial wastewater thus preventing contamination of domestic wastewater (and the sludge) designed for reuse. Table 2 shows the environmental impact of a centralised system versus onsite system.

Table 2 Comparison between environmental impact of centralised and onsite systems

Centralised system	Onsite system
Safe disposal of treated wastewater is primary objective. Nutrients if not removed can cause eutrophication of receiving water bodies. Removal of nutrients and other pollutants have been increasingly required. Reuse of treated wastewater will require an additional piping and pumping system.	Onsite reuse of treated wastewater is usually an objective of onsite systems. Nutrients can be recycled back onto land or separated at source.
Wastewater may be contaminated with industrial wastewater discharged to the sewerage system.	Contamination with toxic substances can be minimised by householder. Treated wastewater and sludge can be applied onto land onsite.
Ingress of stormwater into sewerage system is common, resulting in combined sewerage overflow.	Stormwater can be permeated to groundwater or otherwise managed sustainably (see below under Sustainable Urban Water Management)

There are possibilities for large scale wastewater reuse, for example, treated wastewater from the Woodman Point Wastewater Treatment Plant in Perth for the nearby industrial area in Kwinana. Examples like this one are rare, compared to treated wastewater reuse in the garden for irrigation of plants in place of using water of drinking water quality. It is possible to pipe treated wastewater from a centralised system to an agricultural area, but the energy cost for pumping is likely to be high.

Centralised sewerage systems are also subject to combined sewer overflow during heavy rainfall periods, polluting receiving waters. Industrial wastewaters are commonly discharged to centralised sewerage systems. Contaminants may be difficult to remove from the treated wastewater and from the sludge to make them suitable for agricultural use.

Social

Table 3 summarises the social aspects of centralised and onsite systems. Social aspects include policy issues, institutional arrangements and management, and community attitudes.

Table 3 Comparison of social aspects of centralised and onsite systems

Centralised system	Onsite system
Taken for granted as the norm of wastewater management. Well established government policy, regulatory framework, government agencies or corporation responsible for its management	Fragmented management of septic tanks. Local government responsible for approval of installation. Householders are responsible for management with poor environmental outcome. Rapid development of onsite and decentralised system technology meeting same standard of performance as centralised system, but still with unsatisfactory management. Centralised management is essential.
Considerable investment by governments, and requirement to connect to recoup investment costs.	Investment by householder, and expectation to forego investment when sewerage reaches property.
Community expectation for water reuse in arid regions may force governments to consider decentralised systems.	Environmental sustainability features of onsite and decentralised systems may make communities consider installing them.

Social considerations also include protection of public health, public acceptance of the system including convenience in usage, security and a simple system for management. We believe that social considerations have not been given as high an importance as economics and environmental considerations by professionals in the water industry.

Developments in science and technology have ensured that we can achieve equal public health outcomes from both centralised and small scale and decentralised systems through regular maintenance of decentralised systems. The protection of public health can be achieved for about the same costs in both systems.

Centralised systems, being the norm in the developed world, have been taken for granted by the community. They provide the convenience for the disposal of human wastes, the confidence that the service is there 24 hours each day of the year and that a centralised body is responsible for it. There is in place government policy to ensure the proper operation of the system with the accompanying responsible government agencies and structures to collect the rates (taxes), for example, necessary to ensure that the system operates efficiently.

There is no reason for decentralised systems not to function in the same manner. The current situation is, however, a legacy of fragmented management of onsite systems (basically septic tanks and leach drains). They have to meet the same public health outcome as centralised systems, but the management is largely left to householders. Design and installation is by the house builder based on prescriptive design guides and after approval from local council. There is no further requirement to ensure proper management, until something goes really wrong like toilet overflowing.

The dominance of the centralised system in the developed world is such that when sewerage passes through a property the sewerage rate is due irrespective of whether a connection is made, and there is normally mandatory requirement to connect within a certain period. The basis for this is to ensure economy of scale and hence the success of a centralised system. The latter is important for the investor (usually government or its corporation) to recoup the huge capital cost of a centralised system. In contrast the government does not usually make investment in onsite systems or mandate their maintenance to ensure protection of public health.

In urban redevelopment or renewal where there is main sewerage existing in the area the assumption is that new buildings will be connected to the sewer. An application for exemption from connection is required if a decentralised wastewater treatment system is to be used independent of the sewerage system.

Similarly when urban growth envelops peri-urban areas served by onsite systems, there is also an expectation that onsite systems be abandoned and properties connected to the expanding centralised sewerage system. There may be justification for poorly maintained and functioning septic tanks to be replaced by centralised sewerage and treatment. If however properly designed and maintained onsite or decentralised systems are to be abandoned in preference to connecting to a centralised system, there is the real question of whether it is necessary or desirable. From an economics point of view this is an unjustified imposition to abandon an investment that may not have reached its economic life. Properly designed and operated onsite or decentralised systems achieve the same public health outcome as a centralised system. There are benefits of local water reuse and nutrient recycling. And the question then is the community's choice of whether to retain the onsite and decentralised system in preference to connecting to a centralised system. Environmental considerations would favour retaining the onsite or decentralised system, the economics and public health outcomes being similar.

It has been argued that centralised systems allow for higher density urban development. This is because decentralised systems usually incorporate water and or nutrient recycling, hence land is required for this purpose. On the other hand if a centralised system is required to reuse or recycle water and nutrients then land is required elsewhere for this purpose (plus the additional piping and pumping).

Urban planning therefore plays a crucial role in determining the sustainability of decentralised systems. For example, it is possible to have a high density urban development with a decentralised system, provided that there is land available nearby. This implies the development of what we call 'urban villages' within a conurbation, separated by open spaces. The open land spaces provide the conventional parks and gardens in the conventional town planning scheme, but can also function as land for water and nutrient reuse or recycling.

A community in a peri-urban area may, of course, choose to retain its low density development and maintain its decentralised system, when urban development and the centralised system approaches its border.

Sustainable city

The question of whether we choose a decentralised or centralised wastewater system is therefore closely linked to the question of how we should develop our cities into sustainable cities. This question is also raised in the context of transport planning, air pollution management, energy used for travel, energy used for heating and cooling in buildings, and hence the design of buildings and estates, and the social considerations of personal safety, sense of community and well-being. Many people have pointed out that our cities are unsustainable. Car dependent cities mean that we consume a lot of gasoline (a non renewable resource) and we create air pollution. More roads and highways are constructed with a large land area covered by bitumen (more stormwater run-off, greater surfaces reflecting heat from sunlight). Low residential density means longer distances to travel, and longer pipes to supply water and collect sewerage. Household solid waste has to be collected from larger areas and transported to more distant landfills.

A more sustainable city can consist of higher density urban villages separated by open spaces and interconnected by mass rapid transit (transport) system (Newman and Kenworthy, 1999). An urban village is a walking suburb, very much like a traditional village, where the distance from the centre to the perimeter is not more than a 15 minute walk. For example with a radius of about 1 km and an area of 3 km^2 (300 ha) this village can have between 5,000 people (lower density) and 10,000 people (higher density) or more if desired. Within an urban village, residential density needs not be uniform, and each village can have its unique identity (e.g. architecture). This village can be served by onsite systems (lower density housing) or by one or several decentralised systems (for higher-density housing). Reuse of wastewater (water and nutrients) requires land surrounding the high density housing. The land can be productively used for growing food (urban agriculture), but this may not be feasible or desired by the community. The community may want a more natural forest landscape for the open space, or playing field, or golf course. Local climate may mean that there is excess water in the rainy season, and local topography may dictate different management strategies for parts of the water catchment. Characteristics of a sustainable city are summarised in Table 4.

Table 4 Physical characteristics of a sustainable city

Sector	Characteristics
Water	Water is sourced, used and recycled locally (see Sustainable urban water management, below)
Energy	Energy conservation in place and use of renewable energy; walking (urban) villages connected by mass rapid transit.
Buildings	Use of green architecture to maximise comfort, minimise energy for heating and cooling, use environmentally friendly materials
Wastes	Reuse and recycle all material resources
Biodiversity	Green corridors; grow food locally

On the positive ledger of an urban village there is the possibility of collecting and processing of household solid waste locally. This is particularly the case with the organic fraction (usually the largest fraction by weight), which can be composted or vermicomposted, and then recycled onsite (rather than collecting over longer distances, processing into compost and transporting the product back to household gardens).

In the final analysis the shape and sustainability of a city is (and should be) determined by the community. Our role as water professionals is to provide information on alternatives that are available. In the case of wastewater collection, treatment, disposal and reuse, it is has become clearer that decentralised wastewater systems can have the same advantages as centralised systems, but have a clear advantage in terms of environmental sustainability.

Sustainable urban water management

The debate on centralised versus decentralised systems needs to embrace not only wastewater, but equally importantly water supply and stormwater.

Water is usually imported from long distances to supply a city, even in cases where there is water available onsite. There is no reason why water cannot be collected from roofs (rainwater harvesting) for purposes that do not need very high quality water, although water harvested from roofs can be readily treated for drinking water purposes. When water is locally harvested, the water balance of the locality is better preserved. This means that the water balance elsewhere is not greatly affected (e.g. reduced environmental flow or decreased groundwater levels). Water can, of course, be harvested from the catchment surrounding a high density urban village, thus we have a decentralised water supply system, rather than an onsite system (roof rainwater harvesting).

It is now recognised that stormwater run-off (which is of a better quality than wastewater) should not be mixed with collected wastewater. Stormwater run-off is best treated locally through a series of processes beginning with local infiltration to the ground (permeable pavement, filter strip and swale, soakways and trenches), wetlands and lagoons, with the latter functioning as amenities as well. These help to preserve the local water balance by maintaining local groundwater levels, reducing the severity of flooding, and minimising disturbance to the water balance elsewhere.

These considerations for local water supply and stormwater management also demonstrate that decentralised systems have environmental advantages. The economics of local water supply may not be as attractive as centralised water supply, because the diameter of supply pipes is not generally as large as wastewater pipes and water is supplied at a high pressure. The economics of local stormwater management looks, however, attractive, because stormwater sewerage pipes transporting stormwater run-off to distant disposal sites may not be required. As with the management of decentralised wastewater systems, there is a need to manage decentralised water supply and stormwater centrally (and in the same sense of central control by a government agency, but can be managed by public or private service providers). Again in the final analysis it is up to a community whether its water supply and or stormwater systems should be centralised or decentralised.

What needs to change if a community wants a decentralised wastewater system?

We would like to share our experience with assisting urban estate developers in developing sustainable estates in Perth. The sustainability features in these proposed estates are considered selling points to second or third home buyers, and include passive and active solar building design to minimise energy for heating and cooling, use of environmentally friendly building materials (green architecture), use of renewable energy and incorporation of permaculture landscape, besides sustainable water management. With regard to wastewater management the dominance of the centralised wastewater system described above is a major barrier to its implementation. We believe that the following are required to overcome the barrier for implementation of sustainable urban water management.

Government should formulate policy and a regulatory framework to allow decentralised wastewater systems with centralised management. For urban renewal (whole estate basis) and for peri-urban areas connection to sewer should be voluntary and the sewer rate should not be applied when it can be shown that public health and environmental outcomes are similar to connection to a centralised system. There should be a single government agency where developers can go to obtain a permit to install a decentralised system (a one stop shop). This agency should also regulate service providers maintaining onsite and decentralised systems.

Government should invest in establishing demonstration projects of decentralised systems (in lieu of investment in sewerage and its maintenance) to disseminate information on these systems.

Capacity building of government regulators, developers, their consultants and service providers on decentralised systems and their management is essential.

Conclusions

Onsite, small-scale and decentralised wastewater systems can achieve the same public health outcome as centralised systems and for about the same costs where the service providers in both cases meet operation and maintenance requirements under the same centralised licensing and regulatory framework. Small scale systems can have better environmental sustainability outcomes than centralised systems by recycling or reusing water and nutrients locally.

Widespread use of small scale systems will take place only in the context of urban planning for sustainable cities. One model for a sustainable city is a collection of urban villages of higher population density surrounded by open spaces. In this context small scale water supply, wastewater and stormwater management have environmental sustainability advantages.

Social acceptance of small-scale systems needs the test of time, but demonstration projects to promote their sustainability features should be facilitated by government changing its policy and regulatory framework to allow such systems in the current situation where centralised systems are highly favoured.

References

Biomax (2005), Suresh, *pers. comm.* Perth.

Davis, C. (2004) Onsite and small systems conference. *Water* **31** (3), 32–33.

Hutton, G. and Haller, L. (2004) *Evaluation of the costs and benefits of water and sanitation improvements at the global level.* World Health Organization, Geneva.

National Lifestyle Villages (2005), Owen Henderson, *pers. comm.* Perth.

Newman, P.W.G. and Kenworthy J.R. (1999) *Sustainability and cities: overcoming automobile dependence.* Island Press, Washington D.C.

Spencer, P. (2004) Centralised vs decentralised water systems. *Water* **31** (5), 27–31.

Standards Australia (2000) *On-site domestic-wastewater management.* Australian/New Zealand Standard 1547:2000 Standards Australia and Standards New Zealand.

UNEP (2002) *Environmentally Sound Technologies for Wastewater and Stormwater Management: an International Source Book.* Appendix 1. Health effects associated with wastewater and excreta. United Nations Environment Programme International Environmental Technology Centre, Technical Publications Series no. 20, IWA Publishing, London.

Water Services Association of Australia (2000), WSAA Facts, Water Services Association of Australia, Sydney.

Characterization and optimization of small-scale farming communities for material recovery and sustainable management. The case of Sunchochumo Community in the Peruvian Andes

C.P. Pabón Pereira*, K. Kujawa-Roeleveld*, A.P.J. Mol and G. Zeeman***

*Wageningen University. Sub-department of Environmental Technology. Postbus 8129, 6700 EV Wageningen, The Netherlands. (E-mail: Claudia.Pabon@wur.nl; Katarzyna.Kujawa@wur.nl; Grietje.Zeeman@wur.nl)
**Wageningen University, Sub-department of Environmental Policy, Mansholtlaan 10–12, 6708 PA Wageningen, The Netherlands. (E-mail: Arthur.Mol@wur.nl)

Abstract This paper develops a methodology to systematically analyse and design sustainability scenarios for organic residues reuse, following environmental, economic, institutional, social and technological criteria. From a definition of sustainability, a set of criteria and indicators to assess baseline and proposed scenarios is unfolded in a sustainability framework. A characterization framework in turn proposes the revision of specific characteristics from the context and technological options allowing visualizing the data required for the assessment. The methodology is applied in an exemplary case study carried out in a rural community in Cusco, Peru. Scenarios around small-scale treatment of organic residues in combination with reuse of by-products are constructed and assessed following sustainability criteria. Special emphasis is given to on-site treatment and reuse technologies with the perspective of closing material cycles at the community level.

Keywords Integrated assessment; material flow analysis; multicriteria analysis; organic residues reuse; sustainability; systems analysis

1. Introduction. Towards sustainability assessment of organic residues reuse in rural settings

Rural settings in developing countries where small-scale farming is practised provide an opportunity context for the rational use of resources and the implementation of clean technologies. A residue may be generated close to a process where such residue is needed in the form of a valuable and safer product; poor socio-economic conditions produce awareness on the value of resources and hence of residues; and even ancient cultural values that acknowledge material cycles and holistic nature of life could be also present.

Caution is necessary when promoting promising technological paths for transformation of organic residues. Possible reuse scenarios can also cause environmental problems or social disruption if conditions for their implementation are not carefully studied. Issues like soil pollution with salts or heavy metals, oxygen depletion of surface water, eutrophication and even social acceptability of the practices introduced, can trigger scenarios proposed.

Therefore, in determining the best technological option for the treatment and reuse of organic residues, application of the sustainable development concept is a must. As no general solutions exist, a case-by-case assessment is the recommended approach.

For the study described in this paper, a community in Canchis province in Cusco (Peru) was selected. The province, where 101 communities coexist, is part of the high Andes (3400 and 3800 metres above sea level (m.a.s.l.)) in the Inca region in Peru. This is considered as an extremely poor region where issues of inadequate access to resources and markets, land degradation, scarce water availability and lack of governmental support are combined in an agricultural system that is considered as very low productive at the national level.

Sanitation and water provision developed rapidly in Peru in recent years. Twelve thousand sanitation and drinking water provision projects were financed in rural Peru between 1991 and 1999, driven by the overall global goals of reaching complete coverage in drinking water provision and sanitation by the year 2010 (CEPIS, 2000). In this context of poverty and increased intervention from governmental and non-governmental institutions, aid and technological innovations are taking place rapidly. Careful assessment is needed to widen the spectrum of possibilities and balance potentialities and constraints of technological options.

Types of residue relevant for the study are those organic biodegradable of different water content including those of human origin (urine, faeces, kitchen refuse, ashes, grey wastewater), animal origin (manure, urine, dairy products), plant origin (crop residues).

The Intermediate Technology Development Group (ITDG) in Peru provided the institutional setting for performing the research. ITDG is a charity registered in the United Kingdom, which works directly in four regions of the developing world with particular concentration on Peru, Kenya, Sudan, Zimbabwe, Sri Lanka, Bangladesh and Nepal. They have four key areas of intervention as stated in four major goals: reduce vulnerability (natural disasters, conflicts and environmental degradation); help poor people make a better living (production, processing and marketing); help poor people access basic services (water, food, housing and electricity); and respond to the challenges of new technologies. ITDG has been working in Sunchochumo for 11 years.

2. Methods. The designed sustainability assessment methodology

The present research involved three main stages: theoretical revision, field research and data analysis. Figure 1 presents the pathway followed.

2.1 Sustainability framework

The starting point is the construction of a sustainability framework for the assessment of current and other possible scenarios for organic residues reuse.

Time and scale are to be considered when defining the scope of sustainability analysis. In this case one year was designated as the time frame. This unit was subdivided into two seasons, as clear differences in agricultural and household practices exist in the study area. Scale is set at community level although further subdivision in land use systems is done to consider variability within the system. For environmental goals that consider broader scales (externalities), these are considered as out-flows from the main system in the MFA.

Environmental sustainability refers to the ability of the functions of the environment to sustain the human ways of life. Technological developments in organic residues reuse can impact the way these functions are delivered in many ways.

Complementarily to the impacts expected from technologies, if the products are to be reused, consideration of the nuisances that can be caused by reusing by-products in the agricultural environment should be mentioned. Bockstaller *et al.* (1997) propose a set of agro-ecological indicators to estimate the impact of cultivation practices on the agroecosystem and its environment. Bradjes *et al.* (1996) suggest problems to assess and parameters to consider when analysing environmental impact of manure in a livestock system. Martijn *et al.* (c.d) mention considerations of wastewater reuse for irrigation.

Gibbon and Jakobsson (1999) define economic sustainability as the maintenance of a non-declining level of economic welfare now as in the future, which is derived from the income generated by capital stocks. Same authors define social sustainability as having the means to generate enough wealth for a society to reproduce it self and maintain important social institutions, as well as enable a sense of social cohesion and community for the individuals within the society. From a technological perspective, Snel (2002) proposes institutional elements to assess when considering wastewater reuse.

Apart from the previous considerations, technologies themselves possess attributes regarded as essential for a technological scenario to be considered sustainable. Balkema *et al.* (2002) propose a set of them in an overview of sustainability indicators to compare wastewater treatment systems. Other indicators are suggested by Zeeman and Lens (2003).

Measuring sustainability of technological scenarios implies the consideration of global goals attainment as those briefly described; however, choosing the criteria and indicators to assess sustainability is a subjective choice. Table 2.1 show the selection made for the purpose of this study based on an exhaustive revision with the aim of giving a complete and structured view on sustainability assessment.

Table 2.1 Sustainability framework to assess organic residues reuse scenarios

General objective	Specific objectives	Criteria	Indicators	Measurement (Relevant flows or information)
Maximize environmental performance	Minimize pollution	Groundwater pollution	Nutrient leaching	Total N in flows reaching groundwater (kg N /year)
		Surface water pollution	Nutrient released to water	Total N lost into surface water (kg N/year)
			Organic matter released to water	Total COD in flows reaching surface water (kg/year)
		Soil pollution	Salt soil pollution	Possible risks associated with inorganic salts present in flows reaching productive soils
			Heavy metals pollution	Possible risks associated with heavy metals present in flows reaching productive soils
		Gaseous emissions	Harmful gases released to atmosphere	Release of gases related to acidification and greenhouse effect (CH_4, CO_2, NH_4 NO_x, O_3)
	Maximize efficiency in resource use	Nutrient recovery	Nutrient efficiency	Nutrients reused/ Nutrients available (%)= Total N, of organic residues being reused / Total N in organic residues

General objective	Specific objectives	Criteria	Indicators	Measurement (Relevant flows or information)
Maximize economic performance	Maximize farmer income level	Energy recovery	Net energy production	Energy generated from reuse of residues = Total energy produced from organic residues (m3 CH_4 per household per year) – total energy demanded by the process
		Water recovery	Fraction water reused	Greywater reused /greywater available (%)
		Animal productivity	Valuable feeding production	Amount of protein produced from residues and used to feed cattle (Kg protein/year)
		Crop productivity	Extra fertilization	Total N from organic residues reaching soil in LUS A (kg N/year)
			Extra amendment	Total amendments prepared with organic residues available LUS A (kg per household/year)
	Minimize economical requirements	Financial demands	Investment cost	Land+labor+equipment+experimentation costs ($ per capita)
			Operational and maintenance costs	Labor+materials+indirect costs ($ per capita)
Maximize social benefits	Improve living conditions	Health risks	Pathogen exposure	Risk of pathogen contact from handling residues
			Toxic gasses exposure	Health risk associated with exposure to harmful emissions form organic residues
		Human nutrition	Valuable food production	Extra protein produced from organic residues (kg/year)
	Recognize and enhance the existing social structure	Social viability	Social acceptability	Change of routines and farmers perception implied in the alternative
			Social manageability	Simplicity in social managerial demands of the scenario
		Equity	Benefits and costs distribution	Distribution of benefits and costs among farmers according to socioeconomic categories

General objective	Specific objectives	Criteria	Indicators	Measurement (Relevant flows or information)
Maximize Institutional learning and support	Recognize and enhance the existing institutional structure	Institutional suitability	Institutional feasibility	Possibility of having scientific, NGO and governmental support
			Legal viability	Possible constraints from current legislation
Maximize technological manageability	Manageability and endurance of technology	Technological appropriateness	Robustness	Capacity of the system to overcome fluctuations
			Flexibility/ adaptability	Possibility to extend the system in capacity or with additional treatment

2.2 Characterization framework

When attempting criteria assessment on sustainability grounds is important to state from the beginning the information needed to measure their performance in later stages. Context and technological information is required to assess sustainability criteria.

Context information is the basis for determining the specific environmental risks and problems, the availability and quality of the organic residues, the unsatisfied needs (human and from the farming system), human attitudes and capacities, legal demands, social capital and institutional support (Figure 2.2).

In a parallel analysis, intrinsic characteristics of technological options should also be defined. The objective is to determine the potential of different alternatives in the recovery of valuable elements, the delivery of by products, the expected quality of this by products and the specific requirements for its appropriate functioning (Figure 2.3).

3. The Study Area

3.1 Canchis Province in the Peruvian Andes

Canchis is one of the 13 provinces that form Cusco department. It is located in the upstream watershed of Vilcanota river at an average altitude of 3,548 m.a.s.l. occupying an area of 3,999 km^2. Canchis province comprises seven districts, Sicuani town being the capital of the province.

Peasants are mostly dedicated to the production of Andean products for their own consumption. Farmers' activities vary according to altitude. The existence of natural pastures and the climatic conditions seem to favour livestock production: species of camel, sheep and cattle account for most of the production in the region.

Farmers are organized in communities that are autonomous in their organization, in their communal work, and in the use and free disposition of their lands, as well as in the economic and administrative aspects within the legal framework that the law establishes.

3.2 Sunchochumo Community

Sunchochumo community is a peasant community located 25 km from Sicuani town. Lands are located between 3492.5 and 3875 m.a.s.l., the total extent being 188.75 ha. Farmers identify three main zones in their community: the pampa or valley zone (3500–3490.5

m.a.s.l.) which constitutes approx. 50% of their land; ladera or intermediate zone (3500–3625 m.a.s.l.) occupying approx. 30% of their terrains; and the altura or high altitude zone (3625–3875 m.a.s.l.).

Seventy-one families live in Sunchochumo community. They account for 217 people of whom approx. 80 are children. The population is evenly distributed in three neighbourhoods, which cover especially the pampa zone (78%). These neighbourhoods are of about the same extension, having similar population densities around the average for the community, which is 4.18 inhabitants per ha.

Among the many economic activities that provide farmers with an income are livestock, small animal raising (cuyes, chicken, ducks and pigs) and dairy products. Livestock production is undoubtedly the main income-generating activity, whereas crop production (mainly maize, fava bean and potatoes) is for self-consumption. Three socio-economic categories were identified when accounting for land tenure and livestock ownership.

4. Results

4.1 Potentialities and constraints

Main potentialities and constraints for organic residues reuse in Sunchochumo community are described based in the characterization as well as on the analysis of technological options for the treatment of organic residues, the focus being those providing useful by-products. Technologies considered include: composting, vermicomposting, anaerobic treatment, mushroom production, algae ponds, fish polycultures, ensilage and alkali treatment, sand filters and constructed wetlands. Table 4.1 summarizes potentialities and constraints.

Table 4.1 Potentialities and constraints for organic residues reuse in Sunchochumo

Potentialities	Constraints
Environmental problems and risks	
• Lack of drainage in the pampa, diminish the risk of ground water pollution and infiltration losses • Low drinking water consumption and low level of sanitation: Upgrading in the direction of sustainable water management is possible without major cultural and infrastructure obstacles. • Change latrine system: groundwater pollution, get inundated in rainy season and farmers want to reuse their contents.	• Salinic soils in the pampa can be a constraint for water reuse. • Fluctuating and shallow ground water table in the pampa. • Complexity arises from the different land types and land uses. • Avoid polluting groundwater in ladera which is a source of drinking water if water reuse for irrigation is considered.
Residues availability	
• Different types of residue are produced: cattle manure, cattle urine and cleaning greywater are most appealing fractions to recover in terms of amount, each of them accounting for 2595, 1126, 1166 ton/year, respectively. • Cattle manure already being collected in the dry season, collecting in the rainy season could be profitable. • Concentrated greywater is produced and improperly disposed.	• Careful consideration of residues reuse in order not to interfere with current uses: manure already being used as fuel and for composting and vermicomposting, crop residues and kitchen refuse being used as cattle feeding. • Distances for residues transportation. • Small amount of land owned by each producer

Potentialities	Constraints
	• Residues like household animal manure, kitchen residues, dairy wastewater and ashes are not appealing to recover as they are produced in small fractions and are already being reused.
Needs	
• Signs of surface water pollution and possible ground water contamination are a challenge. • Being livestock the main income delivery activity, improving feeding will certainly deliver more profit than burning them or applying them to land. • Farmers' perception that soil quality needs to be improved. • Irrigation water needed in ladera • Cattle feeding demanded • Human respiratory problems related to manure burning	• Low income and saving capacity • Heavy workload could make farmers reluctant to participate in projects that demand high labour.
Human attitudes and capacities	
• Farmers good perception of material reuse, acknowledgement of material cycles and already existing expertise, for example: current use of human urine for fertilization and amendment preparation, reuse of crop residues to feed cattle, composting and vermicomposting practice and acceptability, and farmers′ recognition on the value of human faeces as soil conditioner • Respiratory problems related to dung burning • Protein demand in farmers' diet. • Gastrointestinal diseases are not prevalent in humans.	• Individualism • Farmers are getting used to receive aid from NGOs this restrict their creativity and create dependence. • Inequalities in knowledge about treatment technologies
Social capital	
• Good organizational arrangements: negotiation capacities, clear rules. • Autonomy and self reliance	• Socio-economic divisions evident in income, land tenure and number of animals, drinking water and sanitation provision.
Institutional roles and perspectives	
• CEPIS research activities in Peru, in the field of wastewater reuse • Expertises in the coast of Peru on water reuse projects. • Presence of various NGOs with local expertise and resources	• Lack of cohesiveness among the institutions working in the region • Public presence not assuming a leadership role.
Legal framework	
• Existence of legislation in the field of waste water reuse • Incentives to private investment in drinking water, sanitation and solid waste initiatives. • Community autonomy recognized by law	• Legal framework for sanitation currently under reformulation. • Secondary treatment is demanded by law when waste water is to be reused for crop irrigation. • Many and costly permits are demanded for wastewater reuse for irrigation.

4.2 Baseline scenario

The Sunchochumo farming system is complex. Natural phenomena impose variation in the system such as the differences between soils in the pampa and in the ladera, or the climatic fluctuation along the year defining dry and rainy seasons. At the same time, variation in farm management practices is also evident as an adaptation to the complexity of the environment that farmers have to cope with, but also as an adaptation to available assets, which is different among the socio-economic categories identified. Social and institutional structures also showed a complex network of relations.

Complexities were analysed in terms of land units and land uses (Driessen and Konijn, 1992). Nine landuse systems (LUS) are identified and further aggregated. The three aggregated land use systems (Figure 4.1) resulting are quantified separately in terms of mass and nitrogen flows using material flow analysis. Solid and liquid organic residues are studied in an integrated approach. Figure 4.2 shows the MFA for LUS A as an example, in Pabon (2004). MFA for the three aggregated LUS can be found as mass flows and nitrogen flows.

4.3 Design and assessment of alternative scenarios

Before drawing technological scenarios two additional steps were necessary: to establish goods priorities (organic residues) and to screen alternatives according to their feasibility.

Main organic residues generated in the community are defined based on the available amount, environmental implications and potential reuse value. As a result solid and liquid cattle manure, human physiological residues and cleaning greywater are the most important.

Technological feasibility is approached by taking into consideration restrictions dealing with climate, legislation and residues availability. For instance, fish polycultures and reed bed filters are hindered by climatic conditions, while composting and vermicomposting are restricted due to the low availability of cattle manure. Stringent legislation on wastewater reuse in Peru raises concern on reed bed filters implementation as well as on urine reuse for fertilization. Anaerobic digestion seems a promising technology if it is able to deliver enough biogas to replace actual manure burning for cooking purposes. Duckweed ponds and sand filters are also appealing technological options.

Finally, two scenarios are constructed and evaluated applying the sustainability framework. The scenarios aggregate technological options in two levels, one showing a house on-site approach, while the other depicts a community on-site one.

The house on-site scenario includes urine diversion toilets, urine reuse in amonification of crop residues and fertilization, anaerobic digestion for faeces and manure treatment. The community on-site scenario involves composting toilet with urine diversion, anaerobic digestion of manure, duckweed ponds for grey water treatment and part of the manure in the rainy season. (Figures 4.3 and 4.4).

As a final step, scenarios are assessed applying the set of indicators shown in Table 2.1. The method employed to evaluate sustainability applies the multi-criteria analysis concept assigning importance to different criteria and indicators from three different perspectives: environmental, local and neutral.

To reach a single sustainability indicator for performance of scenarios a scale from 0 to 3 was used to transform quantitative and qualitative values. In the case of quantitative indicators, linear functions were used. For qualitative indicators, 0 was used for low performance, 1.5 for medium performance and 3 for high performance. The performances were then transformed into percentages taking 3 as 100% (best performance).

The community on-site scenario shows the best sustainability score on all three perspectives (Figure 4.5).

5. Discussion

Sustainability is not to be brought from technologies themselves but its assessment should always be site specific. Sustainable technologies per se do not exist. Local constraints and potentialities coming from the context where a technology is to be implemented makes unrealistic to talk about technologies that will perform well under any circumstances. In addition, the inherent dynamic nature of a technological process and the fact that it is the product of the social context where it is developed and used makes a technology to be always different.

In order to reach a valid assessment of organic residues reuse on sustainability grounds an integrated approach is necessary. Analysing a whole community system allows to visualize integrated solutions that may not be visible when focussing in smaller parts of the system. The method show to be highly demanding in terms of information, some indicators where not quantified for this reason. The question is to what extent reality should be simplified for practical reasons as. In the authors' perspective complexity, inherent to the decision making process, should be addressed.

A methodology was built trying to create a first approach to complexity. However, a long path is to be followed in what refers to the construction of integrated models, adjustable to different situations and easy to interact with. Models are needed to account for the intrinsic complexity of organic residues reuse assessment, make trade-offs visible and make possible the iteration process needed to optimize technological scenarios.

The method uses linear functions to translate indicators value into figures that could lead to the final quantification of scenarios sustainability. Inaccuracy, especially in the case of quantitative data, is present because linear functions are not the best way to make such translation. In addition no thresholds are incorporated into the indicators. This step should be performed in the future, as indicators by themselves do not say anything unless compared to the limits.

Sustainability as a goal can only be conceived if technological alternatives are constructed and reconstructed by local actors, therefore it is a must that scenario options proposed from a theoretical perspective in this paper are discussed and validated in practice with local actors and enriched by further research and empirical experiences.

5. Conclusions

A methodology was proposed to assess sustainability of technological scenarios for organic residues reuse. The methodology, which comprises two frameworks, is then applied in a community case study conducted in Cusco, Peru.

A quantitative assessment of baseline and proposed scenarios is reached using as main tools the MFA and MCA. A Land Use Systems concept is also basic for understanding and simplifying scenarios.

An opportunity context for organic residues reuse was found in Sunchochumo. Solid and liquid cattle manure, cattle urine and cleaning greywater are the most appealing fractions to recover. There is a need for proper treatment and reuse of urine and faeces, provision of irrigation water in the pampa, soil quality improvement, and human health enhancement (especially respiratory problems), among others. Living conditions make upgrading in the direction of sustainable water management possible. Existence of strong cooperation bounds and clear rules of the game as well as good perception of local people on residues reuse and acknowledgement of material cycles are the main social opportunities.

The main constraints found are related to low water availability, salinity of soils, high water table, land and seasonal variability. Also, legislation and institutional frameworks impose restrictions.

Baseline scenarios allowed visualization of three aggregated LUSs in terms of residue availability. Two scenarios are drawn (house on-site and community on-site) and evaluated, being the communal ones the most favoured when applying the sustainability evaluation framework using multicriteria analysis.

6. References

Balkema, A., Preising, H., Otterpohl, R. and Lambert, F. (2002). Indicators for the sustainability assessment of wastewater treatment systems. *Urban Water* **4**, 153–161.

Belevi, H. (2000). Material flow analysis as a strategic planning tool for regional wastewater and solid waste management. Duebendorf: EAWAG-Swiss Federal Institute for Environmental Science and Technology.

Bockstaller, C., Girardin, P. and Van der Werf, H. (1999). Indicators: tools to evaluate the environmental impacts of farming systems. *Journal of Sustainable Agriculture* **34** (4).

Brandtjes, P.J., De Wit, J., van de Meer, H.G. and Van Keulen, H. (1996). Environmental Impact of Animal Manure Management. Wageningen University, The Netherlands. http://www.fao.org/WAIRDOCS/LEAD/X6113E/x6113e00.htm.

CEPIS/OPS/OMS (2002). Evaluación de los Servicios de Agua potable y Saneamiento 2000 en las Américas. http://www.cepis.org.pe/eswww/eva2000/peru/informe/inf-BIB.htm.

Driessen, P.M. and Konijn, N.T. (1992). Land-use systems analysis. Wageningen University, The Netherlands.

Dragun, A and Tisdell, C.(1999). *Sustainable Agriculture and Environment. Globalization and the Impact of Trade Liberalization.* Cheltenham: Edward Elgar Publishing.

Evans, G. (2001). *Biowaste and Biological Waste Treatment.* UK: James and James.

Glave, M. *et al.* (1998). Typology of farmers in the Vilcanota Valley. Sicuani-Lima: ITDG Peru.

Lens, P., Zeeman, G. and Lettinga, G. (2001). *Decentralised Sanitation and Reuse: Concepts, systems and Implementation.* London, UK: IWA Publishing.

Martijn, E, Huibers, F. and van Lier, J. (c.d). Appropriate agricultural use of treated effluent under (semi-)arid climate conditions. Wageningen University, The Netherlands.

Pabon, C. (2004). Characterization and optimization of small scale farming systems for organic residues recovery and sustainable management. The case of Sunchochumo Community in Peruvian Andes. MSc thesis, Environmental Technology and Environmental Policy groups. Wageningen University, The Netherlands.

Smaling, E. (1993). An agroecological framework for integrated nutrient management with special reference to Kenya. Wageningen University, The Netherlands.

Snel, M. (c.d). Re-use of wastewater-its advantages and disadvantages specifically from an institutional and socio-cultural perspective. International Water and Sanitation Centre. Delft, The Netherlands.

Weaver, P., Jansen, L., Grootveld, G., Spiegel, E. and Vergragt, P. (2000). *Sustainable Technology Development.* Sheffield: Greenleaf.

Zeeman, G. and Lens, P. (2003). Biological wastewater treatment. Part I Anaerobic waste water treatment. Lecture notes. Wageningen Univ. Wageningen, The Netherlands.

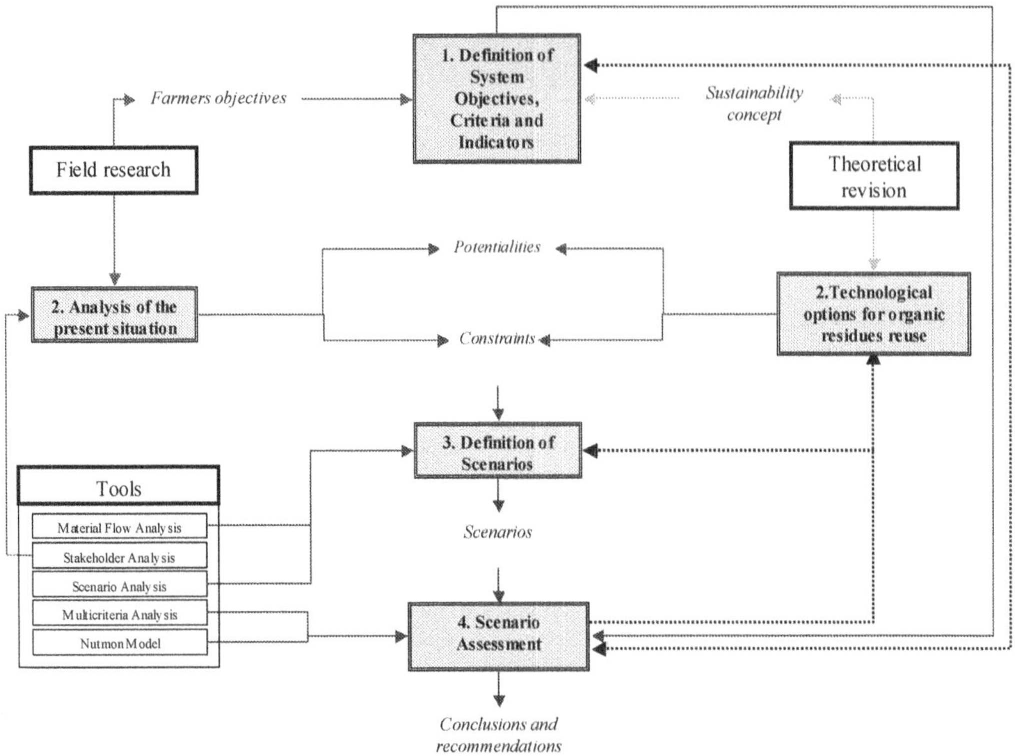

Figure 2.1 Research approach (Pabon, 2004)

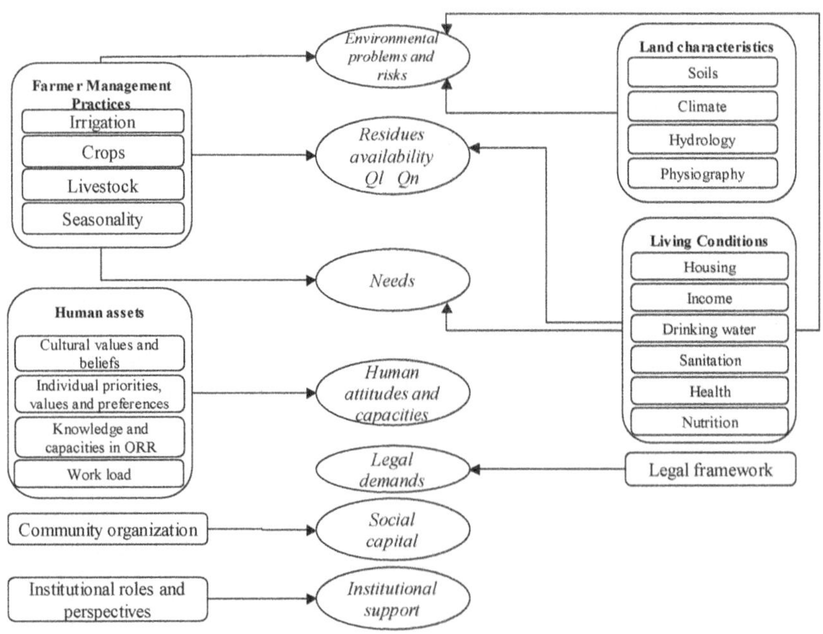

Figure 2.2 Assessment framework for context characteristics (Pabon, 2004)

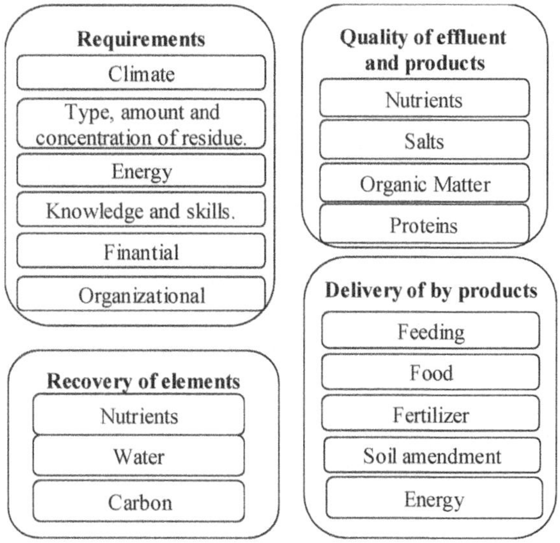

Figure 2.3 Assessment framework for technological characteristics (Pabon, 2004)

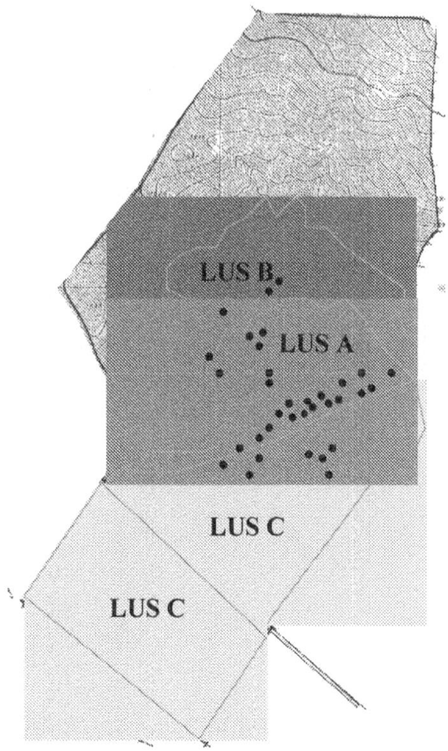

Figure 4.1 Aggregated Land Use Systems for Sunchochumo Community (Pabon, 2004) (LUS A: Crop growing pampa; LUS B: Crop growing ladera; LUS C: Pastures. Dots refer to housing location)

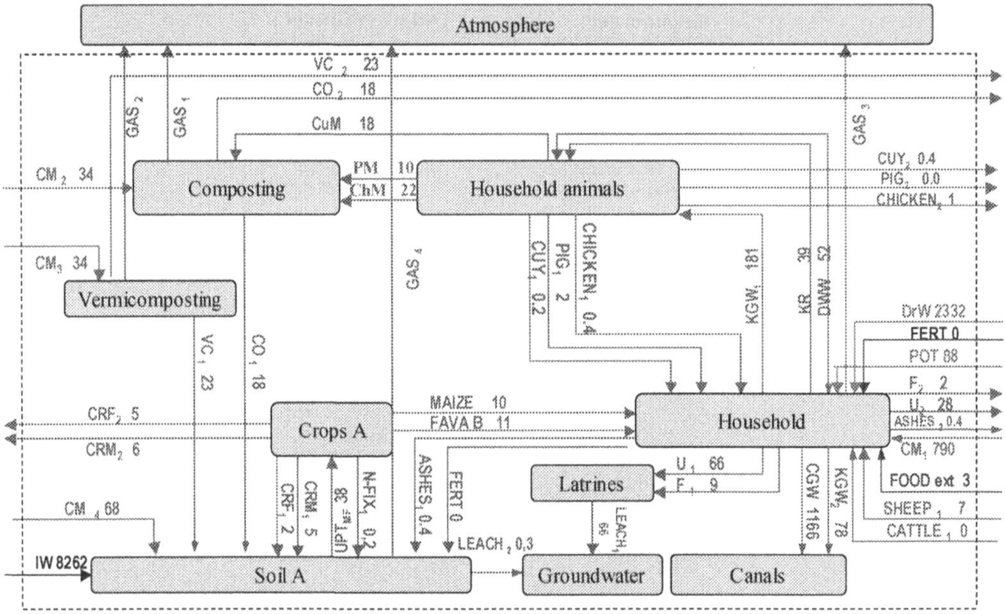

Figure 4.2. Baseline scenario. Quantified Material Flow Analysis for LUS A, Crop growing pampa, in tons per year. (For LUS B and C refer to Pabon, 2004)

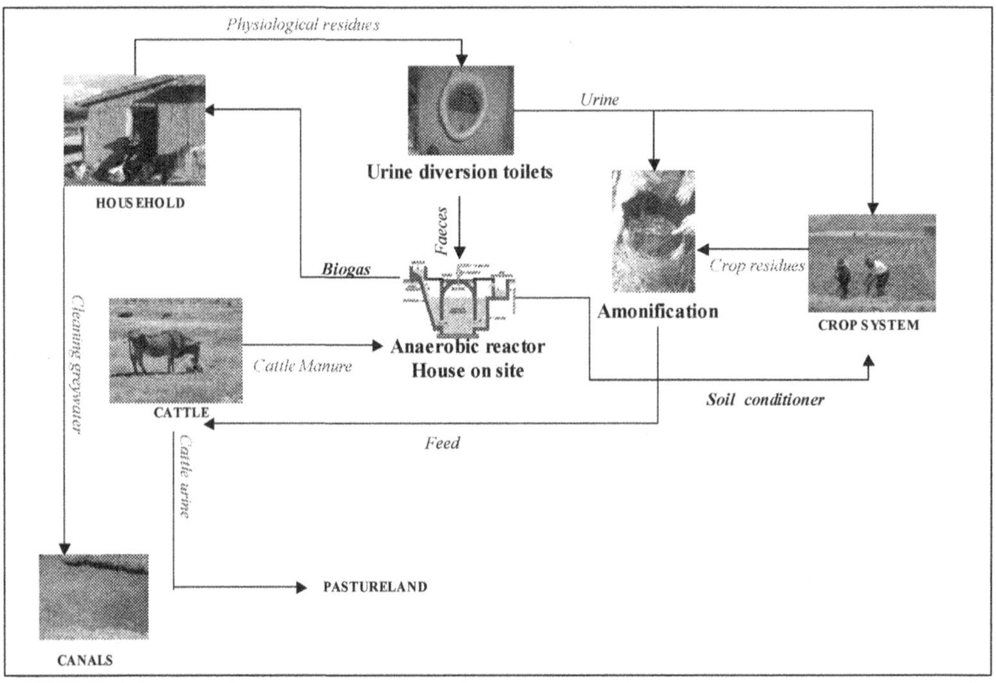

Figure 4.3 House on site scenario for Sunchochumo Community (Pabon, 2004)

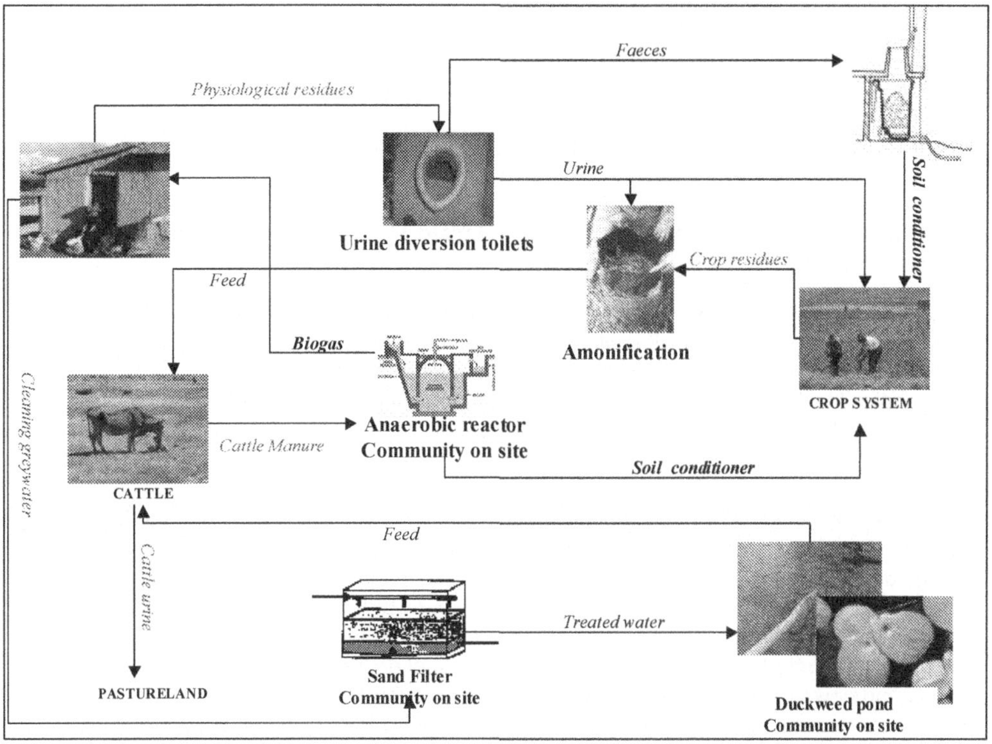

Figure 4.4 Community on site scenario for Sunchochumo Community (Pabon, 2004)

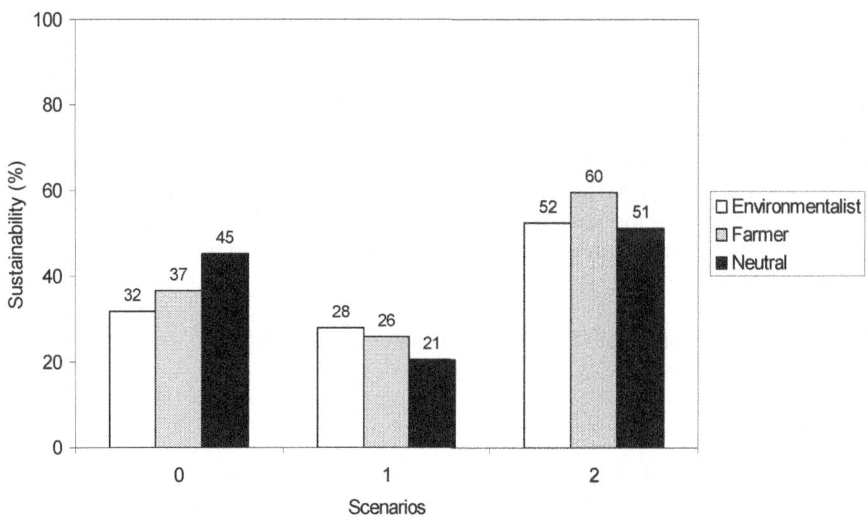

Figure 4.5 Sustainability assessment of scenarios: 0 Baseline, 1 House on-site, 2 Community on-site (Pabon, 2004)

Demonstration-project DEUS 21: a concept for a sustainable urban water infrastructure

Christiane Kotz*, Thomas Hillenbrand*, Harald Hiessl* Marius Mohr and Walter Trösch****

*Fraunhofer Institute Systems and Innovation Research (ISI), Breslauer Strasse 48 D-76139 Karlsruhe, Germany. (E-mail : Christiane.Kotz@isi.fraunhofer.de; Thomas.Hillenbrand@isi.fraunhofer.de; Harald.Hiessl@isi.fraunhofer.de
**Fraunhofer Institute Interfacial Engineering and Biotechnology Stuttgart (IGB) Nobelstrasse 12, D-70569 Stuttgart, Germany (E-mail: Marius.Mohr@igb.fraunhofer.de Walter.Troesch@igb.fraunhofer.de)

Abstract As dynamic as the changes are today in many towns above ground, underground they are often still very conventional: the traditional pipelines and systems of water supply and sewage disposal found today in industrialised countries began their existence in the middle of the nineteenth century. They supply drinking water and water for fire fighting and dispose of polluted water and rainwater. Since it is indisputable that water supply and sewage disposal play a dominant role for each individual and the national economy as a whole, there is the need to find new and sustainable alternatives to today's large-scale centralised structures. Adapted small-scale systems with modular components which use the latest technologies represent a promising solution for sustainable water management. Such systems also have application advantages for newly industrialising and developing countries. This paper will describe the innovative concept for sustainable urban water infrastructure called DEUS 21 which is an acronym for the German "decentralized urban water infrastructure system". The DEUS 21 concept utilises rainwater for water supply, and collects residential wastewater together with biogenic kitchen wastes for joint anaerobic treatment. The concept is due to be implemented in a newly developed residential area with about 100 single family homes in the municipality of Knittlingen, Germany, starting in September 2005. The paper describes the concept of DEUS 21, presents the system dynamics based approach to modelling water and energy flows and balances associated with various water infrastructure concepts and gives an overview of the approach of the sustainability assessment of the DEUS 21 concept in comparison to the conventional approach. The paper also reports on practical aspects related to the planning phase.

Keywords decentralisation, sustainability, urban water infrastructure

Introduction

As dynamic as the changes are today in many towns above ground, underground they are often still very conventional: many of the traditional pipelines and systems of water supply and sewage disposal found today in industrialised countries began their existence in the middle of the nineteenth century. They supply drinking water and water for fire fighting and dispose of polluted water and rainwater. Since it is indisputable that water supply and sewage disposal play a dominant role for each individual and the national economy as a whole, there is the need to find new and sustainable alternatives to today's large-scale centralised structures. Adapted small-scale systems with modular components which use the latest technologies represent a promising solution to sustainable water management. Such systems also have application advantages for newly industrialising and developing countries.

The research project DEUS 21, which is being funded by the German Federal Ministry of Education and Research, and supported by the municipality of Knittlingen, Germany, implements an innovative concept of municipal water infrastructure in a new residential development area for approx. 100 houses. The project aims to design urban water infrastructure in a more sustainable way and to develop alternatives to conventional water supply and wastewater disposal.

Three research institutes and several industrial partners are participating in this project. The Fraunhofer ISI is responsible for the conception, development and application of a computer-based system for modelling the water, energy and nutrient flows in the DEUS 21 residential area. Furthermore the Fraunhofer ISI prepares a database of suitable technologies for modern urban water supply, wastewater infrastructures, and waste disposal in residential areas. In addition, a life cycle assessment as well as an economic assessment of different concepts will be conducted by the Fraunhofer ISI.

The main function of the Fraunhofer IGB is to design the overall infrastructure concept for the DEUS 21 area together with partners from the industry and the city council of Knittlingen. This task includes the projection of the water supply and wastewater treatment facilities in Knittlingen; in more detail, the treatment plant for rainwater runoff from streets, the design of rainwater collection from roofs, the rainwater conditioning and distribution network in the residential area to each private property as well as the planning and building of a semi-central wastewater treatment plant. The Fraunhofer IGB is also responsible for the operation and optimisation of the technical assets.

The Institute for Environmental Engineering (ISA) which is part of the Department of Civil Engineering of the Rhineland-Westphalian Technical University (RWTH) in Aachen is responsible for supplying the nitrogen/phosphorus recycling technology. The operation and optimisation of that equipment is done by the Fraunhofer IGB.

The concept

The DEUS 21 infrastructure concept consists of the following components and technologies.

Rainwater and non-potable water supply

It is useful to permit seepage of rain runoffs rather than discharging these into the sewage system for several reasons including flood protection, accumulation of ground water and for an efficient operation of the treatment plant. Within the framework of DEUS 21, rainwater is handled in an environmentally-friendly way. The rainwater from roofs is collected in a subterranean system of storage drains. This water is treated using a membrane process. Although this water will chemically and hygienically conform to the specifications of the German Drinking Water Ordinance (Trinkwasser-Verordnung TVO), it will be supplied to the households only for uses not requiring potable water quality (e.g. washing and bathing, cleaning, toilet flushing and garden irrigation) via a dual supply system. For legal reasons, potable water (according to German drinking water standards) has to be supplied to each house through the conventional public water supply system. Since the treated rainwater has a very low degree of hardness, it is especially suitable for hot water applications (e.g. for operation of washing machines and dishwashers as well as for showering), makes the use of descaling agents superfluous and reduces the demand for soaps and detergents. The DEUS 21 project represents the first time in Germany that rainwater is used as a resource to provide a dual non-potable water supply for all household applications not explicitly requiring drinking water quality.

Sewage system

A vacuum sewage system is being installed to collect domestic wastewater on the DEUS 21 development area. The building owner has the option to either directly install this system in his house or connect the house to a subterranean transmission station in the conventional way. If the vacuum system is installed in the building, vacuum toilets can be used in bathrooms. These toilets save 80 % of the amount of flushing water used by conventional toilets.

Biogenic kitchen waste

Since the anaerobic wastewater treatment process can process biological kitchen waste alongside domestic wastewater, the houses will be equipped with a food waste or garbage disposer (grinder, macerator) to grind kitchen waste. The ground kitchen waste is then discharged to the vacuum sewer system treatment. The joint treatment of wastewater and kitchen waste has two advantages: (1) in the warm season, the kitchen waste in rubbish bins does not emit unpleasant odours or cause hygienic problems, and (2) due to its relatively high content of fat, oil, grease and other organic compounds, kitchen waste considerably increases the biogas yield if co-treated anaerobically with the residential wastewater.

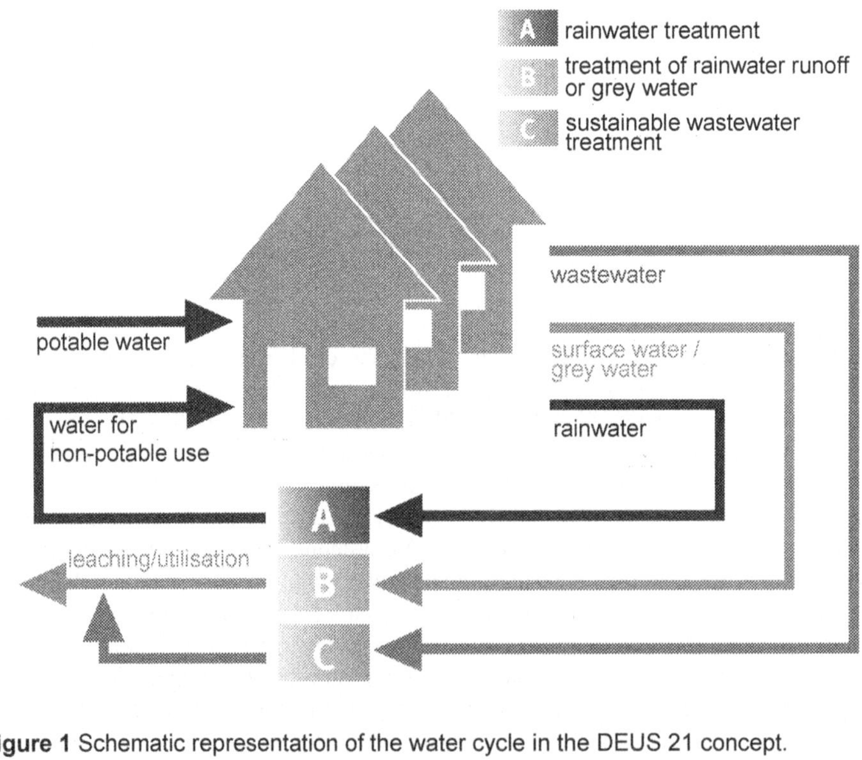

Figure 1 Schematic representation of the water cycle in the DEUS 21 concept.

Wastewater treatment

The Fraunhofer IGB has developed a modern concept for wastewater treatment in urban communities. This concept is a new approach to a decentralised treatment of urban wastewater. It aims to use all wastewater components as a source of material and energy. The concept consists of four modules that are explained in the following:

Module 1: Membrane filtration of wastewater

The fundamental concept behind this process is the separation of the domestic, preferably highly concentrated wastewater flow into a highly polluted concentrate and a less polluted filtrate flow by means of modern membrane techniques. Due to this separation, the wastewater flows can then each be treated in an optimal way and wastewater compounds can be re-extracted: carbon as biogas, nitrogen as ammonium and phosphorus as phosphorus fertiliser. At the Fraunhofer IGB, a rotating disc filter with ceramic membranes was developed. It consists of a cylindrical housing, in which a stack of membrane discs are fixed on a rotating hollow shaft. The discs rotate at 200–800 revolutions per minute during the filtration resulting in a run-off of the surface layer. This system has a low energy demand and works very efficiently in comparison with other filtration processes.

Module 2: Anaerobic decomposition of the highly polluted fraction

The concentrated flow will be metabolised into biogas (methane/carbon dioxide) with the aid of a progressive anaerobic technology. The biogas formed can be used in a combined heat and power plant for the generation of energy, heat or cold.

Module 3: Aerobic membrane bioreactor for the less polluted fraction

The solid-free, less polluted filtrate flow will be transformed into CO_2 and biomass by means of modern wastewater reactors with a highly efficient oxygen supply. This aerobic process step works with high biomass concentrations. The excess biomass can be used to generate biogas in the anaerobic process. The treated wastewater can then be discharged into receiving water or used for other purposes. As technical innovation and alternative to the aerobic treatment an anaerobic process has been developed by Fraunhofer IGB to clean the filtrate at psychrophilic temperatures.

Module 4: Nitrogen and phosphorus recycling

Biomass from sewage sludge will be fermented in the anaerobic digestion system (module 2): the organic carbon is captured in the biogas; bound nitrogen is transformed into ammonium (ammonia). Due to the high concentration, it can be economically recycled out of the sewage sludge, for example by means of stripping systems. This can be realized by an air stripping process in a packed column, leading the air stream in a counter- current to the filtrate and thus removing the ammonia. The air stream is freed of the ammonia in an acid scrubber and used in the stripping column again. The ammonium is accumulated in the acid and can be precipitated and used as a fertiliser in the end. Nitrogen compounds that are contained in the filtrate flow after the first membrane separation (module 1) are also present as ammonium. They can be stripped or precipitated and then recycled as a technical product.

Remaining phosphorus compounds can be removed after the first or second membrane (module 1 or module 3) from the solid-free wastewater, whereas the precipitate has to be recycled as a fertiliser.

In the frame of DEUS 21, the concept of central rainwater utilisation will be implemented for the first time in a newly developed residential area. All the facilities for the wastewater treatment will be constructed on land near the development, where the rainwater treatment plant and the vacuum station will also be installed. Due to the fact that the residential area will be developed gradually, the facilities will also be enlarged step by step to ensure the optimisation of the treatment processes.

Sustainability assessment

Beside the practical implementation of the technical concept, an important part of the project consists of conducting a comprehensive sustainability assessment of the DEUS 21 system compared with a conventional water infrastructure system concept as applied in a neighbouring development area. Due to the long technical lifetime of water infrastructure systems it is insufficient to include only current boundary conditions. The assessment has also to take into account future changes of the environment.

System dynamic modelling approach

In the DEUS 21 project a system dynamics model is used to quantify and analyse the water flows of the infrastructure concepts and the specific technical subsystems applied. In system dynamics, real-world processes are represented in terms of stocks (e.g. stocks of water), flows between these stocks, and information that determines the value of the flows (Forrester 1968). The main assumption is that the internal causal structure of a system determines its dynamic tendencies. Not single decisions or external disturbances are responsible for a system's behaviour, but the structures within which decisions are made (Richardson 1991). Among others, three main factors have an important effect on the amount of water and energy used in a certain infrastructure concept and these are therefore represented in the model.

First, the demographic change. The German Federal Statistical Office assumes in its tenth population projection that the population in Germany will decrease even if positive immigration of 200 000 people takes place each year. At present there are 82.5 million living in Germany, in 2050 this figure will have dropped back to the level of 1963 (approx. 75 million). But it is already conceivable that in various regions, especially in some new Federal States, the decrease will be much more serious.

The second factor is climate change and its consequences which affect the operation and functioning of the urban water and wastewater infrastructure. The dimensioning of the existing combined sewer system was based on static factors. For individual regions of Baden-Wuerttemberg, it can now be shown that short and intense rainfall events are on the increase. Under these new conditions, the combined sewers are no longer big enough. If the sewer system is to be renovated or extended, such unstable factors of climate change have to be taken into account.

Third, the technological change. Figure 3 shows the average development of the water consumption per capita and day in Germany since 1979 (after 1991 including the new Federal States) together with the minimum and maximum values of water consumption in German Federal States. Especially due to the development and installation of innovative, i.e. more water-efficient technologies, for example household appliances like washing machines[1], overall water consumption has been decreasing since 1987. However, the spectrum of water consumption also differs between the Federal States which is attributed to different behavioural patterns of consumers and to various technical equipments in different regions.

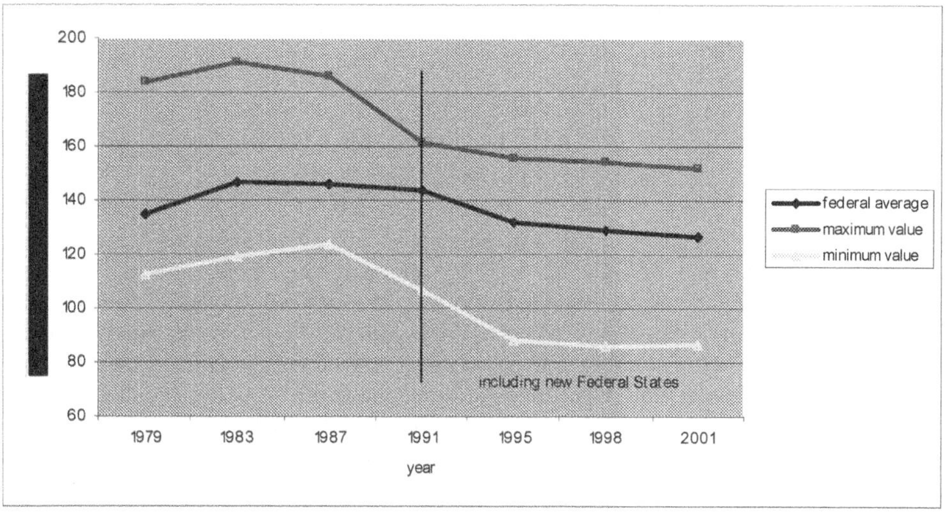

Figure 2 Development of water consumption per capita per day in Germany

Using the system dynamic model makes it possible to analyse the resulting stocks and flows of water and wastewater in the residential area of Knittlingen. By systematically varying the parameters which determine water consumption and wastewater amount, it is possible to model the water and wastewater flows for different alternative urban water infrastructure concepts in the area as well as to identify data for an economic assessment of alternative systems.

Sustainability criteria

On the basis of the outcome of the system dynamic model a sustainability assessment will be conducted. Practical ecological criteria for a sustainability assessment have been proposed by the Federal Environmental Agency (UBA 1999). But the sustainability assessment of water infrastructure systems requires some additional specific assessment criteria to take the particularities of these systems into consideration (e.g. long technical lifetime, spatial extent).

For example, in the project AKWA 2100 (Hiessl et al. 2003), the Fraunhofer ISI assessed the implications of certain water infrastructure concepts for instant on **ecological criteria** like

- aquatic and terrestrial eutrophication
- human and ecological toxicity
- burden on resources (e.g. water, phosphorus)
- greenhouse effect
- acidification

---

[1] The water consumption of washing machines decreased from approx. 180 litres in 1970 to approx. 75 litres in 2000.

on **economical criteria** like

- investment costs
- systemic aspects (e.g. flexibility, expandability of components)
- synergetic effects
- safety aspects (e.g. damage frequency, measure of damage)

and on **social aspects** like

- comfort of handling
- effects on employment
- risk of flood
- transferability to newly industrialised and developing countries.

In DEUS 21, the Fraunhofer ISI will conduct a life cycle assessment taking these sustainability criteria into account as well as an economic assessment of alternative components. A material balance model based on the software UMBERTO is build to calculate the values of sustainability criteria. Two important points in that model are first, that all aspects from water supply up to wastewater as well as rainwater management are included and second, that not only the utilisation phase but also the building phase is integrated, in which differences between various concepts are expected.

Project phases

The first administrative steps have been made successfully, in August 2004 the permit concerning water law has been granted. Currently the infrastructure of the residential area is in construction. In September 2005 the first collected rainwater is expected, so that the technical realisation of the project can start.

Conclusion

Global water problems can only be solved by innovative decentralised water supply and sanitation systems. The concept developed in the DEUS 21 pilot project is not only useful in many industrialised countries, where the sewer systems are in a bad technical condition and urgently need repair and renovation but also for newly industrialising countries and developing countries affected by water shortages, because the construction of a complex and costly conventional sewer systems for a centralised collection of wastewater is not necessary in the DEUS 21 concept.

References

Forrester, J.W. (1968) Principles of Systems. Cambridge.

Hiessl, H., Toussaint, D., Becker, M., Dyrbusch, A., Geisler, S., Herbst, H., Prager, J.U. (2003) Alternativen der kommunalen Wasserversorgung und Abwasserentsorgung AKWA 2100. Bd. 53, Reihe Technik, Wirtschaft und Politik Physica, Heidelberg.

Richardson, G.P. (1991) Feedback Thought in Social Science and Systems Theory. Philadelphia.

Trösch, W. (1999) Neuer Umgang mit Wasser; in: "der gemeinderat", Sonderausgabe "Kommunale Wasserwirtschaft"; August 2004; Eppinger Verlag OHG, Schwäbisch Hall; p. 10f.

UBA - Umweltbundesamt (1999) Bewertung in Ökobilanzen. UBA-Texte 92/99, Umweltbundesamt, Berlin.

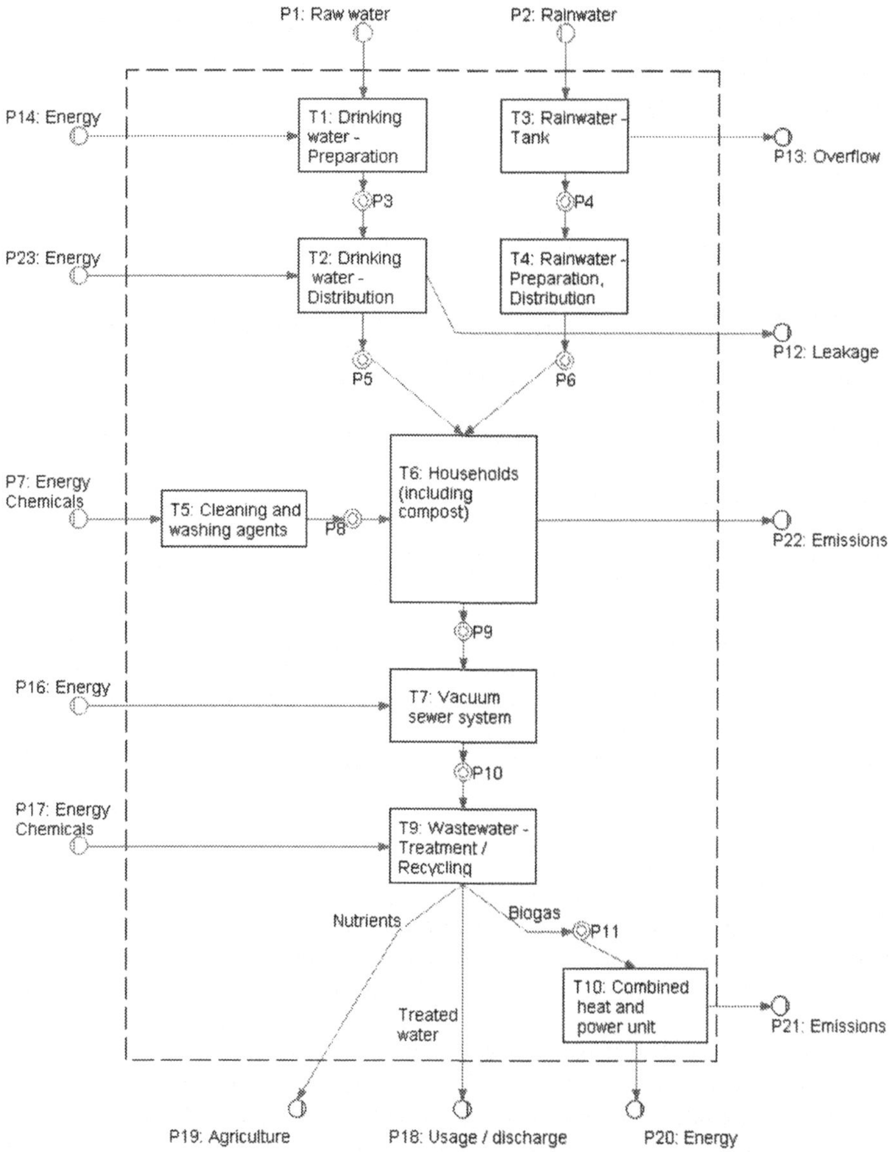

Figure 3 Part of material balance model based on the software UMBERTO

The characteristics of household wastewater and biodegradable solid waste – a proposal for new Swedish design values

B. Vinnerås[*,], H. Palmquist[***], P. Balmér[****] and H. Jönsson[*]**

*Swedish University of Agricultural Science, Department of Biometry and Engineering, P.O. Box 7032, SE-750 07 Uppsala, Sweden. (E-mail bjorn.vinneras@bt.slu.se)
**National Veterinary Institute, SE-751 89 Uppsala, Sweden.
***Luleå University of Technology, Div. of Sanitary Engineering, SE-97187 Luleå, Sweden.
****RYAAB, Karl IX:s väg, SE-41834 Gothenburg, Sweden. Present address Östesnäsvägen 16, SE-431 36 Mölndal, Sweden

Abstract The present Swedish design values (DV) for urine, faeces, greywater and biodegradable solid waste were scrutinised and compared to the composition of these fractions in two blocks of flats, Gebers and Ekoporten. For evaluation of the greywater composition, two additional areas were included, the housing area Vibyåsen and a calculation based upon the composition of the sewage sludge in Ryaverken sewage treatment plant in Gothenburg. The parameters scrutinised were dry mass, wet mass, BOD_7, COD, N, P, K, Cu, Cr, Ni, Zn, Cd, Hg and Pb. Based on the results, new updated Swedish DV were proposed for these parameters in the fractions urine, faeces, greywater and biodegradable solid waste.

Keywords Biodegradable household waste; faeces; greywater; heavy metal; nutrient, urine

Introduction

To evaluate the environmental impacts and to compare different treatment alternatives for household wastewater, the quantities and composition of household wastewater and biodegradable solid household waste need to be known. In conventional wastewater systems, the composition and mass flows of the mixed wastewater are of interest. For source separating wastewater systems, the focus is instead the characteristics of the wastewater fractions; urine, faeces, greywater and biodegradable solid waste.

The present Swedish design values (DV) for the quantities and composition of urine, faeces and greywater were laid down by the Swedish Environmental Protection Agency (SEPA, 1995). The design values for characteristics of urine and faeces were based upon Swedish food consumption data from the early 1990s combined with literature data on human excretion of urine and faeces (SEPA, 1995). In most of the excretion studies, only one fraction at a time was studied. Moreover the measurements only included a small number of people, in most cases below 20 and often below 10, over a short period of time – normally one to ten days. These circumstances make the comparison between different fractions difficult and complicate the task of setting up mass balances for the human body (Blatherwick & Long, 1922; Robinson, 1922; Kehoe *et al.*, 1940; Johnston *et al.*, 1949; Johnston& McMillan, 1952; Trémolières *et al.*, 1961; Lentner *et al.*, 1981).

Most of the excretion studies are presented in Lentner *et al.* (1981), which is an amalgamation of measurements performed mainly between the 1920s and the 1960s. Since the excretion studies were quite old, the values were adapted to the food consumption data from the early 1990s, which made the design values more reliable (SEPA, 1995). However, there was still a need for scrutiny of the data, as the flows of substances change over time – some heavy metals in society are decreasing (Lohm *et al.*, 1997) and patterns of food consumption behaviour are changing. Another drawback with the existing Swedish design values is that no extensive studies have been performed in which all wastewater fractions were included. During recent years, several measurements performed on diverted urine indicated a need for upgrading of the Swedish DV regarding the characteristics of human urine (Jönsson *et al.*, 1997, 2000).

The present Swedish DV for greywater is mainly based on samples from 16 apartments in an eco-village during seven days, complemented with flow analyses in a few other separated wastewater systems (SEPA, 1995). According to Palmquist (2001) and Eriksson *et al.* (2002), there is an urgent need for more information about greywater characteristics in order to evaluate environmental effects and to enable appropriate treatment to be established.

For biodegradable solid waste, a design value for the quantity and composition was proposed by Sonesson and Jönsson (1996) based on several analyses of collected biodegradable solid waste collected in different parts of Sweden.

Objectives

The main objective of this paper was to scrutinise the present Swedish design values for the characteristics of household wastewater fractions (urine, faeces and greywater) and biodegradable solid waste and compare those to more recent measurements on these fractions. The investigation included the parameters: wet and dry mass (WM, DM), biochemical oxygen demand during seven days (BOD_7), chemical oxygen demand (COD_{Cr}) and concentrations of the elements of nitrogen (N), phosphorous (P), potassium (K), copper (Cu), chromium (Cr), nickel (Ni), zinc (Zn), lead (Pb), cadmium (Cd) and mercury (Hg).

Methods

Several investigations of fractionated wastewater were performed to determine the characteristics of the different fractions. Where large discrepancies were revealed, new design values were proposed for the characteristics of household wastewater fractions and biodegradable solid waste. Data on the characteristics of wastewater fractions were collected from four different sources (Table 1). Field measurements were performed at three locations in Sweden: (I) Ekoporten in the city of Norrköping: a block of flats with 35 inhabitants, corresponding to 1225 person days; (II) Gebers in the suburban area Skarpnäck near Stockholm: a block of flats with 81 inhabitants, corresponding to 1701 person days; and (III) the Vibyåsen housing area in Sollentuna with 169 inhabitants, corresponding to 85 person days. The field measurements corresponded to a total amount of 3095 person days. Furthermore, data on greywater characteristics were calculated backwards from sewage sludge from several wastewater treatment plants in the western part of Sweden during several years in the late 1990s (Balmér, 2001). All the chemical analyses were performed at certified laboratories according to authorised Swedish standard methods.

Table 1 Person days investigated at the different sites plus the different fractions analysed or calculated at the sites

Site	Person days	Urine	Faeces	Greywater	Biodegradable Solid Waste
Ekoporten	1225	x	x	x	x
Gebers	1701	x	x	x	x
Vibyåsen	85			x	
Gothenburg	-			x	

Ekoporten

The Ekoporten block of flats consists of 18 flats equipped with double flushed urine-diverting toilets. The sewage system is divided into three sub-systems with separate pipes; one for greywater, one for faecal water (faeces, toilet paper and flushwater) and one for the urine mixture (urine and flushwater). The study was carried out during 35 consecutive days in the spring of 1999, divided into three periods of 11, 12 and 12 days, respectively. The amount of water used for toilet flushing was recorded and the number of flushes was measured by using electronic counters according to the method described by Jönsson *et al.* (1997). During the study periods, the residents recorded the time they spent at home on recording sheets.

During the study, the urine mixture from all of the inhabitants was collected in a 1 m^3 plastic tank and mixed and sampled at the end of each period following the method described by Höglund *et al.* (2000). The faecal fraction consisted of faeces, toilet paper and flushwater. The sampling of the faecal fraction was performed every day, which is more exhaustively described in Vinnerås & Jönsson (2002).

The samples of greywater were collected proportionally to the flow. The greywater was conveyed to a 1 m^3 plastic tank. When the volume exceeded 0.700 m^3, 0.60±0.01 m^3 was automatically pumped out of the tank, regulated by two inductive switches sensitive to the water level. From a mixed sample of a bleed stream at the outlet, 10 grams of sample were collected per 0.60 m^3.

The solid biodegradable waste was collected in a plastic bin placed in front of the normal disposal point and emptied once every second day. After weighing, the material was emptied onto a tarpaulin, mixed and made into a pile. The pile was then divided into eight equal portions, from which two opposite portions were removed. The procedure of making a pile and removing two eighths was repeated until only 10% of the waste was left. The remaining biodegradable waste was then stored at -20°C until it was ground in a mincer with 12 mm holes. After grinding, the period samples were mixed and triplicate sub-samples were taken for analysis.

Gebers

Gebers is a converted hospital that now consists of 32 flats with 81 residents and is situated in the suburban area Skarpnäck nearby Stockholm, Sweden. The feature of the wastewater system is source separation into the four fractions urine (sparse use of flushwater), faeces (dry toilets), greywater, and biodegradable solid waste. The sampling of source separated urine, faeces, greywater and biodegradable solid waste from households in Gebers was performed during three weeks in October 2001, divided into three periods of one week each. All collected samples were stored at -20°C until analysis.

The urine was collected in three 25-litre plastic containers, which were emptied manually twice a day and the urine was subsequently mixed to a weekly mixture in a 1 m^3 tank. The sampling of the urine was performed according to Höglund *et al.* (2000).

The faeces, which were collected in separate bins for each apartment, were emptied after each of the three sampling weeks. The faecal matter and the used toilet paper were mixed to a homogeneous paste after addition of deionised water and the samples were then extracted.

The samples of greywater were collected proportionally to the flow by equipment for automatic sampling. For every 100 litres of greywater passing a ski board, 160 ml of sample was collected. Every day the collected greywater from the two sampling points was pooled into daily samples. After 7 days, weekly samples were pooled from the daily samples.

The biodegradable solid waste was collected in plastic bins located next to the communal compost reactor. The bin was sampled twice a week, and after weighing the biodegradable waste was emptied onto a tarpaulin where it was mixed and made into a pile. The pile was then divided into eight equally sized portions, from which two opposite portions were collected. The collected biodegradable waste was then treated as the samples in Ekoporten (above).

Vibyåsen (greywater only)

The sampling was performed in the housing area Vibyåsen in Sollentuna, Sweden. The housing area consists of one-family houses comprising 47 households. Each of the four sampling occasions lasted for three hours. Two of the samplings occurred at *low flow periods* (two weekdays between 10:10-13:10 and 12:30-15:30, respectively) and two at *high flow periods* (one Wednesday 17:00-20:00 and one Sunday 11:40-14:40).

The samples were taken in a well where the greywater from all of the houses passed through. At sampling, the well outlet was dammed up and the samples were collected. Samples were collected every 6 minutes and subsequently pooled to a 3-hour mixed sample. The greywater flow in Vibyåsen was measured manually in conjunction with the sampling. The greywater was collected in a 1500 ml beaker while measuring the time with a stopwatch. The collected greywater was decanted into a larger beaker (5000 ml). When almost full, the total volume was read off for the total time. This procedure of flow measurements provided the average flow during the 3-hour sampling occasions.

Calculation from sewage sludge to greywater

The wastewater arriving at a municipal sewage treatment plant has its origins in different types of sources; households, stormwater, landfill leachate and industrial wastewater. At the Gothenburg regional wastewater treatment plant (Sweden), the sources of heavy metals have been studied for several years (Balmér, 2001). The per capita contribution of heavy metals from households was calculated from the sludge of the 10-percentile treatment plants with no assumed stormwater or industrial discharges. The per capita amount of sludge was assumed to be 60 g DM p^{-1} d^{-1} and the distribution of heavy metals ending up in the sludge at the wastewater treatment plant was calculated according to Balmér (2001). The heavy metal contributions to wastewater from households obtained this way agree with the data found in the household wastewater fractions described above. If the heavy metals in urine and faeces are deducted, the difference represents the greywater contribution.

Results and Discussion

All of the field studies are first presented and discussed separately; thereafter the composition of the wastewater fractions is discussed and compared to the present Swedish design values. Where significant discrepancies were revealed, new design values for household wastewater fractions and solid biodegradable waste were proposed. The values for urine and faeces were presented so that the excreted amounts corresponded to the amounts expected if 24 hours a day were spent at one place. This assumption was possible since the distribution of urine and faeces appears to occur consistently distributed throughout the day (Hellström & Kärrman, 1996; Pharmacia, 2000). However, the amounts of greywater and solid biodegradable waste were assumed to be produced to 100% at home since those fractions arise from everyday household activities that mainly occur at home. In the attempt to present generally applicable data, the composition of the tapwater was deducted from all the results.

The flow of substances in society is reflected in the wastewater and the solid wastes. As the habits and consumption patterns change over time, the composition of waste fractions will be affected, and therefore such design values have to be updated in the course of time.

Ekoporten

The average time spent at home in Ekoporten was 13.9 hours per day (58%) during the sampling periods. In the double flushed urine diverting toilets, some of the urine is misplaced and collected in the faecal bowl, giving lower amounts of nutrients collected in the urine and larger amounts in the faeces. This is also reported in several other studies on urine-diverting toilets (Jönsson *et al.*, 1997, 2000).

To correct for the amount of misplaced urine, a mathematical revision of the measured values from Ekoporten was made. Approximately 32% (Vinnerås & Jönsson, 2002) of the urine nutrients was therefore deducted from the faecal fraction and added to the urine fraction. Both the adjusted values (AV) and the measured values (MV) are presented (Table 2). The N:P-ratio of the urine mixture was 9 instead of the expected 11-12 according to Höglund *et al.* (2000), possibly due to nitrogen losses when handling the samples.

Compared to earlier measurements and present DV, high amounts of heavy metals were detected in the urine (Table 2) and the faeces (Table 3) in Ekoporten. The main source was most likely heavy metals in misplaced domestic cleaning water that the residents poured into the toilet due to the lack of a suitable sink in the households. Heavy metals are found in dust from households, which consequently end up in the domestic cleaning water (Koch & Rotard, 2000). In the adjusted values the surplus amounts of heavy metals in the two toilet fractions were added to the greywater as a precaution to make sure that the collected metals were charged to the right fraction. Due to uncertainties in the exact pollution sources, these surplus metal concentrations were also retained in the toilet fractions and thereby included both in the toilet fractions and in the greywater. However the surplus of copper in the urine was not added to the greywater since the major copper source was identified as being the half metre of copper pipe connecting the toilet to the urine soil pipe in the house (Vinnerås *et al.*, 1999).

Gebers

The average time spent at home in Gebers was around 15 h p^{-1} d^{-1} (63%) during the sampling period. The mass flows of urine and urine mixture (urine + flushwater) were on average 1.39

and 1.77 l p^{-1} d^{-1}. The N:P ratio of the urine was on average 15, which was higher than the expected 11 (Höglund *et al.*, 2000). This may be a result of phosphate precipitation, e.g. calcium phosphate, in the pipes (Vinnerås *et al.*, 1999). The metal concentrations in the urine were adjusted for the flushwater to avoid local influence from the tapwater. The largest metal influence from flushwater was for Ni (12%). The others ranged from 0-5%.

The load of Cr, Ni and Zn in the faeces was quite high; this may be a result of the galvanized soil pipes connecting the toilets to the faecal collection bins. The DM of the faeces was 13% (10.0 kg p^{-1} y^{-1}), which was only one third of the design value. The total amount of DM in faecal matter (faeces and toilet paper) was 18.5 kg $p^{-1}y^{-1}$ with a water content of 77%. These figures were in general agreement with other investigations presenting 23% DM for faeces and toilet paper (Vinnerås & Jönsson, 2002). The comparatively high wet mass of the faecal fraction may originate from misplaced urine. The total mass flow of greywater, 40 m^3 p^{-1} y^{-1}, was measured by the tapwater consumption, minus the flushwater (for urine). The greywater dry mass (DM) was 14.6 kg p^{-1} y^{-1} of which 99% was volatile solids i.e. organic matter. At Gebers the average greywater BOD/COD ratio was 0.45.

Vibyåsen (greywater only)

As expected, large variations in the greywater flow during the different sampling periods were observed, as a consequence of the residents being at home or not and due to different activities over the day, e.g. morning showers and dishwashing after dinner. At the high flow periods, the average flow was about 15 l min^{-1} and at the low flow periods about 3 l min^{-1}. An approximated average daily greywater flow at Vibyåsen was 66 l p^{-1} d^{-1}, based upon 60% high flow periods and 40% low flow periods daily. At Vibyåsen the average greywater BOD/COD ratio was 0.71 corresponding to the average concentrations 418 mg l^{-1} (BOD_7) and 588 mg l^{-1} (COD_{Cr}) which were about 50% and 25% higher, respectively, than in the Gebers greywater. Large variations in oxygen demanding substances occur in greywater (Eriksson *et al.*, 2002) and the Vibyåsen and Gebers values were on a par with the greywater concentrations of BOD (90-360 mg l^{-1}) and COD (13-8,000 mg l^{-1}) reported by Eriksson *et al.* (2002). The ratios of BOD/COD for untreated municipal wastewater typically range from 0.3 to 0.8 (Tchobanoglous *et al.*, 2003). The measurements at Vibyåsen are reported in Palmquist (2001) in more detail.

Urine

The collected amounts of urine were 554 kg p^{-1} y^{-1} at Ekoporten and 510 kg p^{-1} y^{-1} at Gebers, deducting the flushwater volumes (Table 2). Hellström and Kärrman (1996) reported the collected urine amounts from 30 persons during 24 hours to 550 kg p^{-1} y^{-1}. In several studies by Jönsson *et al.* (2000) where urine mixture was collected during four to six weeks for 8 to 60 persons, with a total of 6500 person days, the average urine amounts were reported to 550±75 kg p^{-1} y^{-1}. The current Swedish design value (DV) of urine wet mass is 365 kg p^{-1} y^{-1}. We propose an increase of the DV for the urine wet mass to 550 kg p^{-1} y^{-1}.

Although discrepancies between measured values and the existing DV were revealed, no change of the DV for dry matter content in urine was proposed. The present DV of the urine dry mass was based upon excretion data – i.e. the urine characteristics directly after release. Since the Ekoporten and Gebers urine had been transported in pipes and stored before sampling, biochemical degradation of the urine dry matter (mainly urea) occurred during collection. The urea is degraded to ammonia and carbon dioxide (Vinnerås *et al.*, 1999), leaving one third of the dry matter in the urine. The DV indicates what leaves the human body, so no suggestion was made to change the existing DV.

Table 2 Amounts of **urine** collected in terms of mass, nutrients and heavy metals in the studies at Ekoporten and Gebers. Both measured (MV) and adjusted values (AV) for misplaced urine (per person and year) are given and compared to the Swedish DV (SEPA, 1995). Where appropriate, a new DV is proposed

Parameter		Ekoporten		Gebers	Swedish	Proposed
		MV	AV	MV	DV	new DV
Wet mass	kg	376	554	510	365	**550**
Dry mass	kg	7	10	7.0	21	21
BOD_7	g	-	-	1829	-	-
COD_{Cr}	g	-	-	3723	-	-
N	g	2100	3700	3830	4000	4000
P	g	230	340	250	365	365
K	g	800	1190	820	910	**1000**
Cu	mg	1490	-	17.2	37	37
Cr	mg	10.3	-	0.16	3.7	3.7
Ni	mg	32.7	-	3.65	2.6	2.6
Zn	mg	150	-	102	16.4	16.4
Pb	mg	15.3	-	4.2	0.73	0.73
Cd	mg	0.48	-	0.08	0.37	**0.25**
Hg	mg	0.30	-	0.16	1.1	**0.30**

Nutrients

Regarding the contents of nitrogen and phosphorus in urine, no significant differences were revealed, either in Gebers or in Ekoporten, compared to the present DV. However, the amount of potassium in the Ekoporten urine mixture was 30% above the existing DV. Jönsson *et al.* (1997; 2000) also showed higher amounts of potassium in urine compared to the Swedish DV. The main reason for increased K values may be a larger consumption of potassium, e.g. increased use of potassium-supplemented salt. Thus there are several indications for increasing the design value for potassium to 1000 g p^{-1} y^{-1} (Table 2). The increased DV was also supported by the physiological distribution of K, with >70% in urine and <30% in faeces (Berger, 1960).

Heavy metals

The heavy metals can be divided into two separate groups, essential and non-essential elements. In this case Cu, Cr, Ni and Zn are essential whilst Pb, Cd and Hg are not. The essential elements are present in the diet as well as in mineral supplements and their presence in urine and faeces is thereby influenced by the diet. At Ekoporten, the amounts of the essential heavy metals were considerably higher compared to the design values, while the amounts in Gebers were comparable to the DV (Table 2). The elevated amounts at Ekoporten were most likely due to misplaced domestic cleaning water (poured into the toilets) and corrosion of pipes (Vinnerås, 2001). At both sites investigated, the detected amounts of zinc were almost ten times higher than the design value. This is probably partly caused by contamination from various building materials, e.g. pipework.

Decreasing amounts of the investigated non-essential heavy metals in Swedish society (Lohm *et al.*, 1997; Balmér, 2001) have probably also affected the dietary intake of these metals. However, the detected amount of lead in the urine in Ekoporten was considerably higher than the DV. On the other hand, the levels of lead in all of the wastewater fractions analysed in Ekoporten were higher than expected. This may be caused by deposition of lead from a nearby airfield. The level of lead in the Gebers urine was comparable to the DV

(Table 2). The cadmium amount in the urine at Gebers was only one fifth of the DV, while the amount at Ekoporten was comparable to the DV (Table 2). For mercury, the results from both sites rendered values considerably lower than the DV, which may be interpreted as a decreased flow of mercury in society.

Even though elevated concentrations were revealed at Ekoporten, no increases of the DV were proposed for Cu, Cr, Ni, Zn, and Pb in urine. The reason for these high levels is not known, and some levels were even too high to be related to the urine alone (WHO, 1995). The high values were assumed to be mainly caused by pollution after the excretion. However, the measured levels lower than the DV of certain metals were interpreted as a decrease in the urine flow of the specific metals. Therefore DV for Cd and Hg are proposed to be lowered to 0.25 mg p^{-1} y^{-1} and 0.30 mg p^{-1} y^{-1}, respectively (Table 2).

Faeces

The dry matter (DM) of faeces and toilet paper collected at Ekoporten was 12.5 kg p^{-1} y^{-1} (Table 3). This was significantly lower compared to the present DV of 12.8 kg p^{-1} y^{-1} of faeces and 8.5 kg p^{-1} y^{-1} of toilet paper (Anonymous, 1994). At Gebers, the DM of the faecal fraction was 10.0 kg plus 8.5 kg of toilet paper (Table 3). In a 24 h collection of faecal matter from seven males, Trémolières *et al.* (1961) collected 7.7 kg $p^{-1}y^{-1}$ of dry matter, which is significantly less than the Swedish DV based upon 24 persons (12.8 kg $p^{-1}y^{-1}$) (Pimparkar *et al.*, 1961). The DV for DM of faeces was proposed to be decreased to 11.0 kg per person and year (Table 3). The wet mass of the faecal fraction at Gebers (72 kg $p^{-1}y^{-1}$) was twice the present DV, which corresponds to a water content of 88%. The existing DV for faecal wet mass corresponds to a water content of 63%. Lentner *et al.* (1981) reported a water content of 77% in faeces of 20 adults. An adjustment of the DV was proposed to 51 kg $p^{-1}y^{-1}$.

Nutrients

The content of N, P, K collected in faeces at Ekoporten and Gebers did not significantly differ from the DV (Table 3), so no changes were proposed. The only significant divergent value was the elevated amount of potassium at Ekoporten.

Heavy metals

The amounts of essential heavy metals in faeces at Ekoporten were somewhat above the present DV (Table 3), which was most probably explained by external sources such as e.g. domestic cleaning water. At Gebers the content of essential heavy metals in the faeces was comparable to the DV, except for zinc, which was significantly elevated (Table 3). A probable source was the zinc galvanised soil pipes between the toilets and the collection bins.

The amounts of lead in faeces at both Ekoporten and Gebers were significantly above the present design value (Table 3). The reason for these high levels is not known, but the levels are too high to be related to the faeces alone (WHO, 1995). As the faecal fraction passes through a sewer before collection, the possibility of external contamination cannot be excluded. Therefore, it is proposed that the present design values should be kept unchanged for Cu, Cr, Ni, Zn, Pb and Cd. However, more thorough investigations of the flow of these elements in urine and faeces is needed to confirm more reliable design values. In spite of this, the levels of mercury in the faeces at both sites were significantly lower than the present design value (Table 3), which was based upon a measurement from 10 persons (Skare & Engqvist, 1992). A change of the DV for mercury in the faeces to 3.0 mg p^{-1} y^{-1} was proposed (Table 3).

Table 3 Amounts of **faeces** collected in terms of mass, nutrients and heavy metals in the studies at Ekoporten and Gebers. Both measured (MV) and adjusted (AV) values (per person and year) are given for Ekoporten. For Gebers AV and the amount of toilet paper (TP) used are given. The values are compared to the Swedish DV for faeces (SEPA, 1995) flushwater and toilet paper (Anonymous, 1994). Where appropriate, a new DV is proposed

Parameter		Ekoporten		Gebers		Swedish	Flush	Toilet	Proposed
		MV	AV	AV	TP	DV	water	paper	new DV
Wet mass	kg	18700[a]	18700[a]	72[b]	8.9	36.5	18300	8.9	**51**
Dry mass	kg	12.6	12.6	10.0[b]	8.5	12.8	-	8.5	**11**
BOD_7	g	-	-	1223[c]	-	-	-	-	
COD_{Cr}	g	-	-	1668[c]	-	-	-	-	
N	g	1800	630	710	-	550	-	-	550
P	g	234	126	250	-	183	-	-	183
K	g	920	540	280	<0.9	365	-	-	365
Cu	mg	1060	-	628	-	400	-	-	400
Cr	mg	68	-	47	-	7.3	-	-	7.3
Ni	mg	110	-	81	-	27	-	-	27
Zn	mg	4860	-	16900	-	3900	-	-	3900
Pb	mg	460	-	13	-	7.3	-	-	7.3
Cd	mg	6.4	-	5.7	-	3.7	-	-	3.7
Hg	mg	2.8	-	3.2	-	23	-	-	**3.0**

[a] flushwater included;
[b] faeces only, toilet paper (TP) excluded;
[c] including TP

Greywater

The greywater embodies the largest volume and the highest amount of dry matter when the urine, faeces and greywater are compared. The greywater flows at the three sites were 38 m^3 p^{-1} y^{-1} at Ekoporten, 40 m^3 p^{-1} y^{-1} at Gebers and 24 m^3 p^{-1} y^{-1} at Vibyåsen. These flows were all considerably lower than the present Swedish DV of 55 m^3 p^{-1} y^{-1} (SEPA, 1995). The proposed new design value for the greywater flow was 36.5 m^3 p^{-1} y^{-1}. The DV for dry matter, BOD_7 and COD_{Cr} were proposed to be changed according to Table 4.

Table 4 Amounts of **greywater** collected in terms of mass, oxygen demand, nutrients and heavy metals per person and year at the four sites compared to the present Swedish DV (SEPA, 1995). Where appropriate, a new DV is proposed

Parameter		Ekoporten		Gebers	Viby-åsen	Gothen-burg	Swedish DV	Proposed DV
		MV	AV					
Wet mass	kg	38000	-	40150	24236		55000	**36500**
Dry mass	kg	21.6	-	14.6	15.1		29.2	**20**
BOD_7	g	-	-	7700	10100		10220	**9500**
COD_{Cr}	g	-	-	17500	14250		26280	**19000**
N	g	613	-	510	230		365	**500**
P	g	162	-	220	180		110	**190**
K	g	1450	-	350	200		180	**365**
Cu	mg	3010	3650	2365	1497	2740	2190	**2500**
Cr	mg	360	430	149	91	400	1830	**365**
Ni	mg	300	430	87	266	590	1100	**450**
Zn	mg	4800	6500	2252	1570	3290	18250	**3650**
Pb	mg	515	990	87	62	390	1095	**365**
Cd	mg	11	14	5.1	2.2	17	219	**12**
Hg	mg	3.8	3.8	1.1	0.4	<1	22	**1.5**

Nutrients

At Ekoporten and Gebers the nitrogen amounts in greywater were higher than the present design value, but lower at Vibyåsen. No pertinent explanation for the difference between the sites and the present design value was found. The DV for nitrogen and phosphorus were proposed to be changed to 500 g p^{-1} y^{-1} and 190 g p^{-1} y^{-1}, respectively. The content of potassium in greywater at the three sites was also higher compared to the design value (Table 4). The reason for this was probably an increased use of potassium, for example in liquid detergents and as a mineral supplement in table salt. Therefore we proposed an increased design value for potassium of 365 g per person and year (Table 4).

Heavy metals

The major proportion of the heavy metal load to household wastewater was found in the greywater fraction. However all heavy metal values measured (in the investigations presented in this paper) in greywater were significantly lower than the present Swedish design value, except for copper (Table 4). The metals present in tapwater were deducted from the greywater concentrations for two main reasons (1) to isolate the contribution of metals from household activities and (2) to avoid the local influence of the tapwater. This operation showed that the tapwater was the source of 60% of the Ni, 10% of the Zn and 0-7% of the other metals investigated. The present Swedish design values for greywater were mainly based upon field measurements of sixteen apartments at the eco-village Tuggelite in Sweden during seven days (SEPA, 1995), corresponding to approximately 400 person days. Data presented on greywater characteristics in this paper correspond to approximately 3050 person days. Moreover the calculated data from municipal sewage sludge were based upon 605,000 persons connected to the wastewater treatment plant Ryaverket in Gothenburg, Sweden, for ten years. Since discrepancies were revealed between the present design values and the new greywater data for all of the metals studied, changes of the design values for metals were proposed according to Table 4.

Copper was the only metal proposed to be increased. The concentration of Cu in greywater is most sensitive to local conditions such as corrosion of the water distribution pipes and service pipes. However the new greywater data indicate a higher level of Cu (2500 mg p^{-1} y^{-1}) than the present Swedish DV. Data from our studies show that chromium, nickel and zinc values were significantly below the present DV. The new proposed design values for Cr, Ni and Zn were 365, 450, and 3650 mg p^{-1} y^{-1}, respectively (Table 4).

The levels of lead were in all cases except for Ekoporten considerably lower than the present DV. A possible reason for the high levels of lead in the wastewater at Ekoporten was the nearby propeller plane airfield, which still uses leaded fuel. The new DV for lead was proposed to be 365 mg p^{-1} y^{-1} (Table 4). The levels of cadmium and mercury were significantly below the present design value in all of the new greywater data. For cadmium, the proposed DV was 12 mg p^{-1} y^{-1} and for mercury 1.5 mg p^{-1} y^{-1} (Table 4).

Biodegradable solids

As a nutrient-containing organic fraction from households, biodegradable solid waste was also included in the study. During recent years there has been a growing interest in Sweden in introducing the biodegradable solid waste into the wastewater by implementing solid waste disposal units into household kitchens (Kärrman *et al.*, 2001). At Ekoporten and Gebers, some significant differences in the composition of the collected biodegradable solids were observed compared to the expected amounts. The Swedish DV for biodegradable solid waste was based upon several studies where the biodegradable solid wastes were collected at different kinds of Swedish housing areas (large scale collection). This means that the present design values were based on a larger data quantity than was presented on biodegradable solid waste in these studies. Moreover, the collection strategy for biodegradable solid waste has effects on the composition of the collected material since e.g. local composting gives a less contaminated quality of biodegradable solid waste (Sonesson & Jönsson, 1996).

The analysed levels of cadmium and mercury were considerably lower than the design values (Table 5). This was probably due to the local recycling of the biodegradable solid waste and a general decreasing flow of Cd and Hg in Swedish society. The proposed changes of the DV were for Zn (700 mg p^{-1} y^{-1}), Cd (2.7 mg p^{-1} y^{-1}) and Hg (0.25mg p^{-1} y^{-1}) (Table 5).

Table 5 Amounts of **biodegradable solid waste** collected in terms of wet mass, dry mass, nitrogen, phosphorus and potassium in the studies in Ekoporten and Gebers (per person and year) compared to the Swedish design values (Sonesson & Jönsson, 1996). Where appropriate, a new DV is proposed

Parameter		Ekoporten	Gebers	Swedish DV	Proposed new DV
Wet mass	kg	82.5	66.5	80.3	80.3
Dry mass	kg	21.5	16.5	27.5	27.5
N	g	570	324	550	550
P	g	90	57	104	104
K	g	200	182	82	82
Cu	mg	306	162	549	549
Cr	mg	263	182	137	137
Ni	mg	33.4	89.4	82.3	82.3
Zn	mg	660	675	3570	**700**
Pb	mg	203	40.2	275	275
Cd	mg	2.7	2.48	8.2	**2.7**
Hg	mg	0.21	0.24	2.7	**0.25**

The proposal for new design values

After scrutinising the present Swedish design values for the characteristics of household wastewater fractions by comparing them to a number of recent measurements, a new proposal for the design values of wastewater fractions and biodegradable solids was put forward (Table 6). The values proposed to be changed are written in bold; the Swedish design values from 1995 are presented in Tables 2 to 5.

Table 6 Proposed DV for composition of the different fractions of household wastewater and biodegradable solid waste per person and year, proposed changes are given in bold text

	Unit	Urine	Faeces	Toilet paper	Greywater	Biodeg waste
Wet mass	kg	**550**	**51**	8.9	**36500**	80.3
Dry mass	kg	21	**11**	8.5	**20**	27.5
BOD_7	g	-	-	-	**9500**	-
COD_{Cr}	g	-	-	-	**19000**	-
N	g	4000	550	-	**500**	550
P	g	365	183	-	**190**	104
K	g	**1000**	365	-	**365**	82
Cu	mg	37	400	-	**2500**	549
Cr	mg	3.7	7.3	-	**365**	137
Ni	mg	2.6	27	-	**450**	82.3
Zn	mg	16.4	3900	-	**3650**	**700**
Pb	mg	0.73	7.3	-	**350**	275
Cd	mg	**0.25**	3.7	-	**12**	**2.7**
Hg	mg	**0.30**	**3.0**	-	**1.5**	**0.25**

Acknowledgements

This study was financed by MISTRA and the MISTRA programme Sustainable Urban Water Management, The Swedish Water & Wastewater Association, The Stockholm Water Company, The Sollentuna Water Company, Norrköpings Hyresbostäder, and the Gothenburg Regional Sewage Works (GRYAAB). We greatly acknowledge Ms J. Weglin, Ms Å. Andersson, and Ms. A. Jensen for their hard work during the several weeks of sampling.

References

Anonymous. (1994). Per capita consumtion of paper. European Tissue Symposium Henkilö 1993.

Balmér, P. (2001) Possibilities to improve the quality of wastewater sludges. Water Sciences & Technology 44:11 pp 19-26.

Berger E.Y. (1960). Intestinal absorption and excretion. In: Comar C.L. and Bronner F.(eds) Mineral Metabolism. New York: Academic Press pp 249-286.

Blatherwick N.R. & Long L. (1922). The utilization of calcium and phosphorus of vegetables by man. Journal Biological Chemistry 2, pp. 125-132.

Elfström Broo A., Berghult B. & Hedberg T. (2000). Drinking Water and Corrosion - A Guide for the Water Works, 2nd Edition. VA-forsk report 2000:12 in Swedish. Stockholm:VA-forsk

Eriksson E., Auffarth K., Henze M. & Ledin A. (2002). Characteristics of grey wastewater. Urban Water 4, pp. 85-104.

Hellström D. & Kärrman E. (1996). Nitrogen and phosphorus in fresh and stored urine. Proceedings from the 2nd International Conference on Ecological Engineering for Wastewater Treatment, September 18-22 1995. Waedenswil: Environmental Research Forum pp 221-226.

Höglund C., Vinnerås B., Stenström T.A. & Jönsson H. (2000). Variation of chemical and microbial parameters in collection and storage tanks for source separated human urine. Journal of Environmental Science and Health (A) 35, pp. 1463-1475.

Johnston F.A., Frenchman R. & Boroughs E.D. (1949). The iron metabolism of young women on two levels of intake. Journal of Nutr. 38, pp. 479-489.

Johnston F.A. & McMillan T.J. (1952). The amount of nitrogen retained by 6 young women on an intake of approximately 70 gm of protein a day. Journal of Nutr. 47, pp. 425-433.

Jönsson H., Stenström T.A., Svensson J. & Sundin A. (1997). Source separated urine-nutrient and heavy metal content, water saving and faecal contamination. Wat. Sci. Tech. 35, pp. 145-152.

Jönsson H., Vinnerås B., Höglund C., Stenström T.A., Dalhammar G. & Kirchmann H. (2000). Källsorterad humanurin i kretslopp. VA-FORSK Rapport 2000:1. Stockholm:VAV.

Kärrman E., Olofsson M., Persson B., Sander A. & Åberg H. (2001). Köksavfallskvarnar en teknik för uthållig resursanvändning? En förstudie i Göteborg. VA-Forsk rapport 2001:2. Stockholm:VAV.

Kehoe R.A., Cholak J. & Story R.V. (1940). A spectrochemical study of the normal ranges of concentration of certain trace metals in biological materials. Journal of Nutr. 19, pp. 579-592.

Koch M. & Rotard W. (2000). On the Contribution of Background Sources to the Heavy Metals Content of Municipal Sewage Sludge. 1st World Water Congress of IWA – Conference preprint Book 1 – Industrial Wastewater Treatment. Paris: IWA pp 367-374.

Lentner C., Lentner C. & Wink A. (1981). Units of Measurement, Body Fluids, Composition of the Body, Nutrition. Geigy Scientific tables. Basle.Ciba-Geigy.

Levlin E. Tideström H. Kapilashrami S. Stark K. & Hultman B. (2001). Slam kvalitet och trender i slamhantering VA-forsk report 2001-05. Stockholm: VAV.

Lohm U. Bergbäck B., Hedbrant J., Jonsson A., Svidén J., Sörme L. Östlund C. (1997). Databasen Stockhome - Flöden och ackumulation av metaller i Stockholms teknosfär. Tema V Rapport 25. Linköping: Linköpings universitet.

Palmquist H. (2001). Hazardous Substances in Wastewater Systems - a Delicate Issue for Wastewater Management. Licentiate Thesis 2001:65. Luleå: Div. of Sanitary Engineering, LTU.

Pharmacia (2000). Svenska Folkets Tarmvanor - Den största undersökningen någonsin om magen, dess skötsel och våra attityder till det alra mest privata. Stockholm.Pharmacia Upjohn.

Pimparkar B, Tulsky E, Kalser M & Bockus H. (1961). Correlation of radioactive and chemical faecal fat determination in various malabsorption syndromes. Am J Med 30, pp. 927-939.

Robinson C.S. (1922). The hydrogen ion concentration of human feces. J Biol Chem 3, pp. 445-467.
SEPA (1995). Vad innehåller avlopp från hushåll? NV rapport 4425. Stockholm: NV.
Skare & Engqvist (1992). Amalgamfyllningar en källa till tungmetallexponering. Läkartidningen 89.
Sonesson U. & Jönsson H. (1996). Urban Biodegradable Waste Amount and Composition - Case study Uppsala. Report 201. Uppsala: SLU Department of Agricultural Engineering.
Tchobanoglous G, Burton FL, Stensel HD. (2003). Wastewater Engineering. Treatment and Reuse. 4th Ed. (Int. Ed). Metcalf & Eddy, McGraw-Hill, New York, 181-191 (1819 pp)
Trémolières J., Bonfilis S., Carré L. & Sautier C. (1961). Une méthode d'étude de la digistibilité chez l'homme, le fécalogramme. Nutr Dieta Eur Rev Nutr Diet 3, pp. 281-289.
Vinnerås B. (2001). Faecal Separation and Urine Diversion for Nutrient Management of Household Biodegradable Waste and Wastewater. Report 245. Uppsala: SLU Department of Ag. Eng.
Vinnerås B., Höglund C., Jönsson H. & Stenström T.A. (1999). Characterisation of sludge in urine separating sewerage systems. In: Proceedings of the 4th International Conference – Managing the Wastewater Resource Ecological Engineering for Wastewater Treatment, Norway. June 7-11 1999
Vinnerås B. & Jönsson H. (2002). The potential of faecal separation and urine diversion to recycle plant nutrients in household waste water. Bioresource Technology 82, pp. 275-282.
WHO (1995). Inorganic Lead. Environmental Health Criteria 165. Vammala: WHO.

Synergy in the City: making the sum of the parts more than the whole

C. Mitchell and S Campbell

Institute for Sustainable Futures; University of Technology, Sydney, PO Box 123, Broadway, NSW 2007, Australia. (E-mail: Cynthia.Mitchell@uts.edu.au; Sally.Campbell@uts.edu.au)

Abstract Infrastructure provides the life-blood of our cities, enabling us to congregate in large numbers – it delivers our needs and wants, and removes our wastes.

The pressures on existing infrastructures are significant: demand is beginning to outstrip supply; aging infrastructure poorly maintained presents an increasing risk; and rejection of urban sprawl forces increasing population density. At the same time, the drivers for infrastructure are changing. We are beginning to recognise ecological limits to supply, leading to shifting expectations, for example, from 'remove waste' to 'recapture nutrients'. We now know that a sustainable future requires step changes in material use intensity, which has further infrastructure implications. We have witnessed it already in communications. For water and energy, and therefore, for transport also, the step changes are on the horizon. Community expectations are moving too, for example, from separating home and work towards co-locating them.

In this paper, we explore three responses to these pressures and opportunities. Firstly, we look to ecosystems for principles of sustainable infrastructure. Efficient and effective communication, material and energy flows, and waste removal are also the key to well-functioning ecosystems. We explore what the ecosystem concepts of 'integrity' and 'health' mean when translated into the context of sustainable urban infrastructure. We explain how 'integrity' translates to 'visibility', and 'health' translates to 'resilience', and explore what these might mean in practice. Secondly, we look to the increasingly networked, interlinked nature of cities for opportunities to recognise and design in 'infrastructure leapfrogs': step-change enabling synergies between water, stormwater, wastewater, waste, energy, transport, and communications infrastructures. Finally, we consider tools that will help us identify when and how to act. By definition, forecasting cannot help to identify or plan for the step changes required to respond to all these pressures, since forecasting confines us to incremental change within the boundaries that we know. The combination of backcasting and integrated resource planning might help. Backcasting enables the identification of the kinds of step changes required. Integrated resource planning, incorporating least cost planning, informs the choice and timing of the step changes. It enables logically consistent comparative economic analysis of alternatives because it focuses on meeting the service provided, rather than providing a unit of supply. Together, these three responses could enable city infrastructure to become more than the sum of its parts.

Keywords Systems planning, sustainability, infrastructure, backcasting, water, energy, transport, urban design, integrated resource planning, least cost planning

Urban infrastructure at the crossroads

Physical infrastructure is a necessity for all stationary communities – it is in part what enables us to move away from a nomadic existence and begin to congregate in ever-increasing numbers. Roman aqueducts are part of the much famed early generations of infrastructure, mobile telephones have revolutionized the here and now, and the future is guided by our imagination. Over decades and centuries, the complex network of urban water, energy, transport, and communications infrastructure has evolved and expanded. Now, however, we are at a crossroads. The pressures on existing infrastructure are significant and mounting, and the objectives we want our infrastructure to meet are changing radically.

Infrastructure faces significant pressures

The pressures on urban infrastructure planners are many-fold. The responses are also diverse.

Firstly, existing highly centralised systems are reaching their limits[1] in terms of meeting current demands. Meanwhile, demand for the services supplied by these infrastructures continues to increase as populations[1] and economies grow. Responding to this pressure provides an opportunity to both manage demand down as well as identify alternative supplies (see Mitchell *et al.*, 2004a).

Secondly, these same infrastructures are, at best, being inadequately maintained (see, for example, Engineers Australia's national infrastructure report card (IEAust, 2001) and, at worst, approaching the end of their useful life (e.g. Sydney Water Corporation's recent tender for retrofitting water mains (SWC, 2004)). We are just beginning to realise the massive looming cost of maintaining and replacing our buried infrastructure. For water and wastewater infrastructure replacement in the USA, some estimates put the cost at USD$1 trillion over the next 20 years (AWWA, 2001, p17). In addition, expenditure on repairs is expected to triple because aging pipes are more prone to breakage[2]. Responding to this pressure presents an opportunity to rethink the highly centralised nature of current infrastructure, and to consider investing locally. For water, this means co-locating sources and end uses, investing in distributed generation and treatment, rather than transport; for energy, investing in distributed generation, rather than centralised generation and distribution; and for transport, investing in access rather than mobility (Berry *et al.*, 2004).

Thirdly, we see increasing rejection of urban sprawl. Over just twenty years (1972 to 1992), with independent travel made possible by the private car, Sydney has grown in area so much that the change is recognisable from space (CSIRO, 1998). Melbourne has also spread in most directions but there the policy response has been earlier and stronger, with Melbourne 2030 embodying Australia's first urban growth boundary; a planning-based regulation limiting the area of land available for the city. More people and less land available means either densities in cities must increase or people must be encouraged to settle in rural and regional areas; both are likely. Responding to the pressures of increasing urbanisation and density provides new opportunities to implement step changes towards a qualitatively more effective kind of service provision.

Finally, and perhaps most significantly, we see changes in the objectives that infrastructure is being asked to meet. We expand on this below.

---

[1] For example potable water supplies in Sydney, Perth and Gold Coast; energy supplies to Sydney's CBD; and road transport corridors in South East Queensland and Melbourne.

[2] Through the combination of around 1,000 new residents each week and decreasing average house size, DIPNR expects 500,000 new homes in the greater Sydney Region over the next 25 years (DIPNR 2004 p6).

[3] Some have argued that these are over-estimates, and that improved asset management (e.g. risk-cost approaches) will lead to more effective targeting of investment. Our view is that asset management will provide some relief, however, the size of the problem will be largely unchanged.

Infrastructure objectives are changing

There are two elements to the changes in infrastructure objectives: changes in our understanding of limits to resource availability, and changing community expectations.

The notion of limits to resource availability gained prominence with the publication in 1972 of 'The Limits to Growth' (Meadows *et al.*). We know that a sustainable future for the all the world's population is predicated on significant step changes in material use intensity, perhaps as much as tenfold reductions (Schmidt-Bleek, 1997) or even higher.

Increasing recognition of limits to resource availability is behind two significant shifts in the water industry. These shifts are in their infancy now, and promise to be significant drivers in the short, medium, and long term. The first is the move from commodity supply to service provision[3] (Mitchell and White, 2003), from a focus on 'providing more' (supply side thinking) to 'using efficiently and appropriately' (demand side thinking). A second shift is from a focus on 'removing waste' to 'recapturing nutrients'[4].

Changing community expectations represent challenges of a different kind. We used to prefer to separate industrial and manufacturing areas from residential areas. Changes in workforce demographics and massive improvements in pollution prevention mean that preference has little standing now. Instead, we are turning to mixed use developments, co-locating work and home, changing the need for transport. In keeping with this, we see movement away from highly stratified cities towards high diversity on a local scale: from maximising transport towards centres of exchange to maximising equitable access towards distributed opportunities for exchange.

Conceptual models and tools help choose between paths

Resource limits and community expectations overlaid on the complex, highly constrained, urban system will result in emergent behaviour. Batten explores emergent behaviour as a property of cities (Batten *et al.* 2000), and notes some collective properties of urban systems can be predicted from our knowledge of individual sub-systems but some (or many) collective properties cannot be predicted (p 58).

Responding to these pressures and changing objectives provides significant drivers for systematically rethinking our urban infrastructure, and specifically, for the water industry to change its direction. Firstly, we can begin to design and implement and evaluate qualitatively different systems, e.g. in terms of technology scale and management models. Secondly, there is the search for new synergies between elements of the water cycle and between the water cycle and other infrastructures. Thirdly, we can look for 'leapfrog' technologies – the mobile telephone equivalents for other infrastructures e.g. avoiding water as a catch-all solvent, or avoiding heat as the source of electrical energy.

---

[3] The move to a service economy has already underpinned major change in other sectors. For example, in manufacturing, the conceptual move to providing a service rather than a product has enabled business expansion at the same time as drastic reductions in waste production and extended product longevity (Professor Dexter Dunphy, ABC Radio, 28 April 2002, Transcript available on-line at http://www.abc.net.au/rn/bigidea/stories/s519567.htmand). In the energy and water industry, sub-contracting specific services to manage demand is a significant growth area (Berry et al, 2004, p29)

[4] Global phosphorus deposits are expected to deplete in the next 50 - 80 years (Birch 1976). The mass of phosphorus that leaves Sydney's sewage treatment plants each year is said to be roughly equivalent to the mass of phosphorus used annually in broad acre cropping in NSW, and therefore represents a significant potential resource.

By definition, step changes and emergent properties such as these are difficult to forecast: so we need different approaches. In the sections that follow we explore three responses. Firstly, we seek insights from relevant conceptual models of comparably complex systems (ecosystems). Secondly, we look for examples of design and technologies that could hold the key to unlocking leapfrog synergies within and between infrastructures. Thirdly, we review two tools that help us identify where (backcasting) and when (least cost planning) to act.

Learning from nature

Ecosystems may present a useful conceptual model to gain insights into city infrastructure. Ecosystems are diverse groups of flora and fauna with complex relationships and stringent requirements. Efficient and effective communication, and intricate material, energy and waste flows, are key to both well-functioning ecosystems and cities. Ecosystems can be characterised by 'health' and 'integrity': James Karr argues that these are qualitatively different concepts, and that both are necessary for ecosystems to function well (Karr 1996). These are foreign concepts for infrastructure, so here we translate them to identify key characteristics for well-functioning city infrastructure. According to the Macquarie Dictionary:

> *Health*/ well being, vitality, resilience
> *Integrity*/ 1. honesty 2. state of being entire, whole, undiminished

What might 'health' translate to for a city's infrastructure?

- Adaptive, resilient systems
- No degradation of the serviced site and areas beyond the site
- Liveable, equitable, vital communities

So, as the objectives for infrastructure change radically to reflect our new understandings of resource availability limits and community expectations, eg from 'remove waste' to 'recapture nutrients', resilience is a key characteristic: flourishing despite changing inputs and expectations.

What might 'integrity' translate to for a city's infrastructure? Perhaps making visible the infrastructure that supports us; making visible our reliance and our impact on the natural environment. Enabling the city to be seen and experienced as a connected, integrated system of inputs and outputs.

So, infrastructure that is well functioning, that meets our changing needs and enables progress towards sustainability might demonstrate resilience and visibility. Each concept warrants further exploration and explanation.

The benefits of resilience

An infrastructure system with 'resilience' is one that can respond to the kinds of technological shifts and pressures common in the twenty-first century. Resilience implies vitality and flexibility, an ability to adapt and change. It contrasts starkly with 'lock-in': a gradual 'carving out' of a particular (infrastructure) path (Batten *et al*, 2000).

Centralised infrastructure is associated with perceived 'lock-in'. One contributing factor is conventional cost estimation processes that focus on the incremental addition to a centralised system, resulting in underestimates of the total system costs to provide the extended capacity. For example, in water services, 'head works charges' incorporate the costs of the added load from additional connections on supply and treatment systems. A resilient approach would use a process like least cost planning[5], that compared whole systems and included all the costs associated with supplying the services through various means.

Resilient infrastructure would be modular, enabling rapid uptake of technological improvements, and more closely matching demand with capacity to supply, thereby maximising the performance of both the infrastructure itself and the capital invested in it. Meeting the demands of increasing population densities through infill and brownfield developments requires retrofits of existing urban infrastructure. The choice is between marginal additions to centralised infrastructure, or using the opportunity to begin to make step changes to more resilient systems.

There is an intriguing relationship between resilience and risk. Here, we consider three related issues: calculating risks, financial risk, and security risk.

Risks are generally assessed as the product of the probability of a particular failure and its consequences (see for example Australian Standard 4360:1999). Highly centralised infrastructure may tend towards lower probability of failure, because of organisational drivers and asset management responses. However, the consequences of failure are high simply because of the degree of concentration. So, when a failure occurs in a centralised system, the consequences can be significant for the environment and/or public health because, for example, the number of people affected, or the volume of wastewater inadequately treated, is large. Distributed systems, on the other hand, may tend towards higher failure rates (at least whilst more effective management models are developed and implemented), but have relatively low consequences. For example, a water treatment 'failure' in 1998 resulted in Sydney Water advising its 4 million or so customers to boil their water prior to use. A water quality problem in a household raintank affects only the handful of people using that supply. Thus, distributed systems may be more resilient.

Security has emerged as an increasing concern for key infrastructures in our post-September 11 world. Twenty years ago, Brian Martin (1984) argued that decreasing society's physical vulnerability was a key strategy for non-violent conflict resolution. He argued that highly centralised utilities (including dams and power stations) provide an obvious weak point open to attack. Designing weak points out of the urban system through reducing resource use and implementing decentralised treatment reduces this risk, and is consistent with distributed, resilient systems.

Finally, in a study of wide-ranging mega-infrastructure projects, Flyvberg (Flyvbjerg *et al*, 2001) showed how the irreversibility of investment decisions (connected in part to 'lock-in' and in part to sunk costs incurred through project commencement) and the scale of investment (which on mega-projects can impact on a country or region's economic development) mean both the likelihood and consequence of risks are high. To avoid or reduce these risks, they identify institutional change as crucial, noting the following four recommendations:

- Performance specification rather than concept design by public authorities;
- Requiring formal risk capital;
- Increased transparency in decision making;
- A clear regulatory regime and policy context; and

These recommendations are consistent with the concepts of resilience (meeting the need) and visibility (integrity). This brings us to a discussion on visibility.

---

[5] See Fane *et al.* 2003 for a discussion of least cost planning analysis to compare supply and demand side options.

Visibility

We see visibility as a two-way relationship between people and infrastructure. Firstly, infrastructures need to respond to people's needs for services (as opposed to resources), and secondly, people need clear signals and information about their resource use in order to be able to make decisions and change behaviours.

Just as Flyvbjerg *et al.* (2001) argue for performance specification to overcome the limitations of deterministic concept design, we argue that a service focus moves us away from optimising the flow of litres of water, electrons or motor vehicles. Instead this perspective increases our understanding at a basic level of the service to people—their needs—so we can optimise the resources to meet the needs. At the heart of this approach is a clear understanding of the interaction between the natural resource and people, for example the services delivered to citizens who use water are hot showers and clean clothes. For the latter at least, water is not the only resource capable of delivering this service.

Finite resources moderate this service potential and so necessitate the other side of the two-way exchange: people responding to natural resource signals. Graham and Marvin (2001) describe cities as a "perpetual flux of infrastructurally mediated flow, movement and exchange" (p.8). They say urban infrastructure, including water treatment systems, streets and transport, telecommunication and energy grids are "mediators between nature, culture and the production of the city" (ibid). This very mediation, long a symbol of order and progress, now offers a means to engage people, linking them with their natural environment and increasing our understanding of the nutrients feeding the lifeblood of cities.

Engaged communities where people can make informed decisions about resource use require the pre-cursor of relationships between people and the resources they use. The current well-intentioned form of 'mediation' of people's interaction with resources via largely invisible infrastructures dis-empowers us, reducing both our understanding of the relationship between our actions and resource consumption; and our ability to reduce resource use should we choose to. Other solutions are possible, and visibility could be a key.

Our understanding of how learning happens, and thus how change is enabled, points to the need for visibility of the infrastructure that supports our urban ecosystem. In his seminal book on communities of practice, Etienne Wenger (p229, 1998) explains how learning cannot be designed: it belongs to the realm of practice and experience. Second generation knowledge management (Philips *et al.*, 2004) recognises the need for personal engagement too – it takes a demand-side approach, and focuses on knowledge production by learners on the job. Finally, Donella Meadows (1998) explains "people can't respond to information they don't have. They can't react effectively to information that is inadequate. They can't achieve goals or targets of which they are not aware. They cannot work towards sustainable development if they have no clear, timely, accurate, visible indicators of sustainable development" (p5). The kind of step changes we envisage will involve radically different infrastructures and will need citizens thus engaged with their cities. What follows is our foray into the future of urban infrastructure.

Critically linked urban infrastructures

Dupay (1991) identifies the networked character of modern urbanism as its single dominant characteristic.

These seemingly distinct urban infrastructure systems (water, energy, transport, communications) are actually dependent systems that rely on each other, and have co-evolved over time, driven by policy, governance and business structures, people's expectations, and changing notions of the environment.

Infrastructure linkages are increasingly apparent and some links provide synergistic opportunities. A simple example is that three per cent of energy usage in the US is for water treatment, supply and wastewater treatment and this doubles if you include the energy used in homes to heat the water (Cohen *et al*, 2004). This means saving water saves energy. It is exactly this 'synergy in the city' that we are searching for.

We propose that backcasting and least cost planning across the infrastructure in cities, at a range of scales, applying the criteria of resilience and visibility, will let us 'tunnel through the cost barriers' (Lovins *et al.*, 2002), minimise resource use, tap synergies and provide better service at lower cost. Far from being a utopian dream, existing processes are already moving in this new direction. In this section, we explore potential contenders for energy and water.

Infrastructure leapfrogs: the mobile phones of water and energy

Resource constraints and community expectations mean we need 'leapfrogs' or step changes in resource use and servicing. Resource conservation will be an essential component, and step changes in technology are also needed to meet the kinds of objectives communities are setting for themselves.

As an incredibly abundant molecule, water has been our universal solvent for millennia, and our universal conveyor for centuries. It is in these uses that the leapfrog is most likely to appear. For example, as a solvent, supercritical[6] carbon dioxide holds intriguing potential.

It is in wide use now as an industrial solvent for processes as diverse as decaffeination in coffee and complex pharmaceutical production. In an intriguing advance, an Australian design about to be paraded at an international design event[7] uses supercritical CO_2 as the solvent in a new dishwasher.

As well as avoiding water use, the design is surprisingly energy efficient. Increasing the pressure of CO_2 from atmospheric to supercritical takes only twice as much energy per unit mass as increasing the temperature of water from 15°C to 60°C. In addition, very little CO_2 may be needed, firstly because supercritical fluids have incredible diffusivity, essentially no surface tension, and terrific solvation potential, and, secondly, because it is recyclable - dropping the pressure slightly changes the CO_2's properties (i.e. the supercritical gas reverts to a liquid) and the dishes' contaminants drop out of solution in a highly concentrated form. The CO_2 would be readily separated and collected, and only a little energy would be needed for the marginal increase in pressure to take it back to the supercritical phase for reuse.

In energy, disappearing fossil fuel sources along with carbon taxes and trading permits are just some of the influences encouraging us to discard cumbersome, inefficient generation processes. Fossil heat and multiple mechanical transformations and massive distribution losses will be superseded by direct (bio)chemical energy transformations. Fuel cells hold great potential here (eg Lovins *et al.*, 2002) and glimpses of other technologies are increasingly prevalent.

Hydrogen is the primary input to run fuel cells. The hydrogen economy prospect has not yet overcome the critical issue of a sustainable energy source to make the hydrogen. At

---

[6] A supercritical fluid is defined as a substance above its critical temperature and pressure. For CO_2, this is above 31°C and around 72 atmospheres. Supercritical fluids exhibit properties of both gases and liquids: like gases, they expand to fill any container, and like liquids, they dissolve materials.

[7] The Buzz, ABC Radio National, 9 Oct 2004.

present, fuel cells rely mostly on natural gas; essentially a transition fuel in terms of sustainable energy[8]. Biological production of hydrogen could hold the key: perhaps microbial fuel cells (MFCs) are the next clean energy source.

Prof Bruce Logan and his colleagues (McCann 2004) are making step changes in effectiveness in their quest to use iron-reducing bacteria with extracellular enzymes that release protons to drive fuel cell electricity production whilst metabolising waste carbon sources. Pennsylvania State University reported a major step for this technology this year when they powered a three milliwatt fan using less than a teaspoon of household wastewater—an interactive demonstration of their achievement in cutting the cost by two-thirds and increasing the electricity output by almost 600%[9].

A further leapfrog is possible through combining these two. Given the highly concentrated nature of the nutrients, this waste stream may be the perfect concentrated organics source for the microbial fuel cell. So, we could have a water free, energy efficient dishwasher using an environmentally friendly, chemically powered cleaning agent as well as distributed, high-efficiency, renewable energy production. Step changes indeed are possible, and the scale of the step changes could be significant. We need tools that will help us decide which step changes to invest in and the order of investment *i.e.* when and how to act.

A toolkit for the future

Tools are needed to structure our thinking and planning, both to take us into this brave new world of technologies we can only just imagine and to extend beyond sub-system (or single infrastructure) conceptual boundaries. This multi-infrastructure assessment is where our analysis must transition—we know this because our natural systems thrive in concert with each other and our infrastructures have co-evolved with each other and are increasingly co-dependent. We propose that some tools that might help us make the leap to this broader scale of analysis. We consider these tools next, and how they might assist engineers to make the most of their role in moving towards sustainability.

Backcasting

In searching for marginal and step change opportunities to enable resilient and visible and interconnected infrastructure, backcasting helps identify the 'what'. It has been applied in seeking to understand various ways to reduce resource use (see, for example, Dreborg 1996 and Robert *et al.*, 2002). The first step in backcasting is to describe a desired future state, most successfully done by picking a point in time that is far enough away to allow creative, innovative ideas to enter the realm of possibility and to avoid being constrained by what we know to be the current situation. A thirty-year time frame is usually workable. Given that desired future state, we then look backwards to the present, identifying alternate interim steps and sets of scenarios that will take us from the present to this desired state.

Having identified the 'what', a range of potentially useful scenarios, we need a methodology that will enable comparison between them, and help to identify which ideas to invest in, and in what order. Enter least cost planning!

---

[8] Saddler *et al.* (2004, p140) identify the long-term limits to Australian natural gas reserves.

[9] See http://www.worldchanging.com/archives/000864.html or http://www.psu.edu/ur/2004/microbfuel.html.

Least Cost Planning

When least cost planning is used in conjunction with backcasting (Fane *et al.*, 2004), we can compare a range of ways to move in the direction of some desirable future from where we are in the present, and understand how much we as a community will pay for the journey.

Confronted with major investment decisions, increasing intensity of infrastructure and challenging multi-objectives, some sectors have turned to least cost planning. Least cost planning is an analytical approach based on a clear definition of the service provided by the resources. This shift in focus, away from volumes of water or units of electricity, raises important questions, most predominantly, 'How can we provide the same service using less resources?'. Secondary questions then arise as to how much the changes cost and where the financial burdens and benefits lie. Water service sectors are leaders in the field in terms of applying these principles in Australia and they are reaping the rewards.

The single resource conservation context in which least cost planning has been applied now has multi-objectives overlaid. Even within the water services sector the volume of water and the quality of water services to people are not the only factors in question. Environmental water quality characteristics are one of the additional parameters to maximise. Subject to locationally specific constraints, minimising sewerage effluent or flattening out peaks in demand can also be robustly targeted using least cost planning.

The opportunity exists now to extend this analysis, to ask questions like 'Where should the next 50,000 homes be located in the Sydney region if we are to deliver services to them at the lowest total long-term cost to the community?'. Least cost planning both levels the analytical playing field to allow balanced consideration of options and ensure we take the whole costs into account. We argue that least cost planning works best when we understand it as just one economic input to decision making. As we explain in the following section, evidence suggests that economic tests alone are not sufficient.

People-Environment Balance

In considering the difficult balance between the economy, people and the environment, we argue that if people and the environment are in balance then the economy can be robust but that the reverse is not necessarily true, *i.e.* a robust economy does not, by default, mean a healthy community and sustainable interactions with the environment. An example is that despite Australia's economic prosperity, a review by Fiona Stanley (Stanley cited in Gleeson, 2004) of physical and mental health indicators finds that "whilst death rates are low and life expectancy is terrific, trends in almost all other outcomes [for children] have got worse". Gleeson describes these outcomes as including birth weight, asthma, diabetes, obesity, intellectual disability, depression, anxiety, behavioural problems, drug use and child abuse. Clearly economic viability is an insufficient measurement.

An engaged populace may also demand different decision-making processes, in terms of both the locus of decision-making and the identity of the decision-makers *e.g.* citizens juries. Others (Carson *et al.*, 2002) have argued that processes involving typical citizens and founded on deliberative, representative and participatory principles, may provide greater certainty for policy makers, and may help to resolve the contentious, value-laden environmental issues we face where deciding on the criteria for choosing between options is the first part of the decision making process (Mitchell *et al.*, 2004b).

Concluding Remarks

Our cities have prospered because of the remarkable achievements of engineers to provide for our material needs at ever increasing densities, complexities, and delivery speeds. But the rules are changing. Our well-hidden infrastructure is facing incredible pressure: demand is outstripping supply, and its maintenance requirements are formidable. At the same time, we are beginning to recognise that there are real limits to resource availability, and our citizens are asking for qualitatively different solutions. What ought we scientists and engineers do in response?

We contend that engineers and scientists have a particular role in sustainable development: we have responsibility for devising, designing, and implementing clever ways to use the 'means' provided by natural resources to meet the 'ends' of providing services to satisfy human demands. The constraints outlined above mean that objective decisions are no longer an appropriate response. Instead, we suggest that engineers have a tremendous opportunity to be the 'honest broker' of technical and scientific knowledge (Mitchell *et al.*, 2004b) in open participatory decision processes; to complement our historical strengths in forecasting with the emerging approach of backcasting; to embrace tools like least cost planning that enable us to simultaneously focus on both efficiency and effectiveness (doing the thing right and doing the right thing); to combine engineering and scientific expertise with other disciplines and understandings to achieve outcomes that cannot be conceived within a single perspective. To enact this new responsibility, we need new models for characterising the built environment infrastructure systems we create: resilience and visibility may be the keys.

References

American Water Works Association (AWWA) (2001) Dawn of the Replacement Era – Reinvesting in Drinking Water Infrastructure, Denver

Batten, D., Bertuglia, C., Martellato, D. and Occelli, S. (eds) (2000) Learning, Innovation and Urban Evolution, Kluwer Academic Publishers, Boston.

Berry, T., Campbell, S., Riedy, C., and White, S. (2004) A New Distributed Infrastructure Services Market – Assessing Australia's Future Market Potential. A discussion paper prepared by the Institute for Sustainable Futures for the Australian Council for Infrastructure Development

Birch C (1976) Confronting the Future. Australia and the world: the next hundred years. Penguin Books, England.

Carson L, White S, Hendriks C, and Palmer J (2002) Community consultation in environmental policy making, *The Drawing Board: An Australian Review of Public Affairs* **3**(1), 1-13

Cohen, R., Nelson, B. and Wolf, G. (2004) Energy Down the Drain – The hidden costs of California's water supply, Natural Resources Defence Council (NRDC) and Pacific Institute, August

Dreborg K (1996) Essence of backcasting, *Futures* **28**(9), 813-828

Energy Australia (2003) Demand Management Standard Offer – Implementation of commercial and industrial energy efficiency initiatives to reduce peak summer load in Brookvale and Dee Why, December.

Fane, S.A., Robinson, D. & White, S.B. (2003) The Use of Levelised Cost in Comparing Supply and Demand Side Options for Water Supply and Wastewater Treatment, *Water Supply* Vol 3 No 3 pp 185-192, IWA Publishing.

Fane, S, A Turner, and C Mitchell (2004) The secr*et al*ife of water systems: Least cost planning beyond demand management. *2nd IWA Leading-Edge Conference on Sustainability in Water-Limited Environments*. Sydney, 8-10 November

Flyvbjerg, B., Bruzelius, N. and Rothengatter, W. (2003) Megaprojects and Risk – An anatomy of ambition, Cambridge University Press, Cambridge.

Graetz, R. D., Fisher, R. P., Wilson, M. A., and Campbell, S.K. (1998) Looking Back – The changing face of the Australian continent, 1972 to 1992, CSIRO Available online at: http://www.eoc.csiro.au/lb/astart.htm

Gleeson, B., (2004) The Future of Australia's Cities: Making Space for Hope, Professorial Lecture at Griffith University, February.

Graham, S and Marvin, S (2001) Splintering Urbanism – Networked infrastructures, technological mobilities and the urban condition, Routledge, London.

Institution of Engineers, Australia (IEAust) (2001) 2001 Australian Infrastructure Report Card, Alliance Partners. Available on-line at http://www.ieaust.org.au/library/institution_pubs.html

Integral Energy (2004) Network 2014 – Maximising Sustainable Value in Our Assets,

Lovins A B, E K Datta, T Feiler, K R Rabago, J N Swisher, A Lehmann, and K Wicker (2002) *Small is Profitable*, Rocky Mountain Institute

Karr J R (1996) 'Ecological integrity and ecological health are not the same', in P C Schulze (ed) *Engineering within ecological constraints,* National Academy of Engineering, p97-110

Knowles, Hon C (MP) (2004) Metropolitan Strategy Launch , 22 April, Sydney

Martin, B. (1984) *Uprooting War*. Freedom Press, London.

McCann B (2004) Electricity from Wastewater – an economical advance *Water21*, August, p38-39

Meadows D (1998) Indicators and Information Systems for Sustainable Development: A Report to the Balaton Group

Meadows, D. H., Meadows, D. L., Randers, J., and Behrens, W. W. (1972) The Limits to Growth, Universe Books, New York, 1972

Mitchell, C. and White, S. (2003) Forecasting and Backcasting for Sustainable Urban Water Futures, *Water* Vol 30 No 5, August 2003.

Mitchell, C A, Turner A, Cordell D, Fane S, and White S (2004a) Water conservation is dead: long live water conservation. *2nd IWA Leading-Edge Conference on Sustainability in Water-Limited Environments*. Sydney, 8-10 November

Mitchell C A, Carew A, and Clift R (2004b) 'The role of the professional engineer and scientist in sustainable development', Chapter 2, in A Azapagic, S Perdan, and R Clift (eds) *Sustainable Development in Practice*. John Wiley & Sons, London.

Phillips C, W Allen, M Kilvingto (2004) Is knowledge management the answer for ICM? A proper learning process can influence community decisions. *Water* **31**(3), 63-66

Robert K-H (2002) Strategic sustainable development – selection, design and synergies of applied tools. *Journal of Cleaner Production* **10**, 197-214

Saddler, H., Diesendorf, M. and Denniss, R. (2004) A Clean Energy Future for Australia, Clean Energy Future Group and WWF

Schmidt-Bleek F (1997) MIPS and factor 10 for a sustainable and profitable economy. Wuppertal Institute, Wuppertal.

Sydney Water Corporation (2004) Tender 510021867 Water and Sewer Mains Renewal Program (WSRP) Request for Proposals.

Wenger E (1998) *Communities of Practice: Learning, Meaning, and Identity.* Cambridge University Press

Advanced sanitation with the vacuum truck collection system of human excreta

S. Matsui*, H. Harada*, H. Utsumi*, T. Matsuda* and Y. Shimizu**

*Graduate School of Global Environmental Studies, Kyoto University, Japan
**Research Center for Environmental Quality Control, Graduate School of Engineering, Kyoto University, Japan

Abstract In developing countries, especially in peri-urban areas, there are needs for affordable alternative sanitation systems that may combine treatment of human excreta with resource recovery; such systems may be employed in lieu of the traditional sewage systems. Non-pipe sanitation systems may be a feasible alternative to the traditional pipe-system because the high-cost of developing such systems are derived from the construction of the pipe infrastructure. Japan has a sophisticated non-pipe sanitation system that employs vacuum truck collection from pit-toilets and which has dramatically contributed to improved sanitary conditions in Japan. The Japanese sanitation system is introduced in this paper. Considering that the contents of urine and faeces are very different, separation of urine from faeces leads to efficient treatment and nutrient recovery. This urine-faeces separation methodology may contribute to the development of alternative sanitation systems that may be useful in developing countries. Herein, we propose a vacuum truck-collection system combined with urine-faeces separation as an advanced sanitation system. The advantages of such as system include the following: (1) installation costs are not as expensive as traditional sewage systems, (2) treatment of human excreta is combined with efficient resource recovery, (3) valuable water resources are conserved, (4) salt concentrations in treatment plant effluent are reduced due to urine separation. This advanced system may be helpful in contributing to developing solutions for the combined treatment of human excreta with resource recovery in peri-urban areas of developing countries.

Keywords human excreta, Japan, night-soil treatment, nutrient recovery, urine-faeces separation, vacuum truck

Introduction

Good sanitation facilities to treat human excreta are necessary to maintain good sanitation since pathogens found in human excreta are major causes of varieties of water-born illness. It was stated that the number of people without sanitation access should be reduced by one-half by 2015 in the United Nations Millennium Development Goals (MDGs).

Although attempts have been made to develop sewage systems in urban and peri-urban areas, only about 600 million people currently have modern sewage systems, with treatment plants, available to them (OECD, 2002). Most peri-urban areas in developing countries will not be able to afford sewage systems by 2015. Alternatives to traditional centralised sewage systems are necessary if the sanitation goals of the MDGs are to be met.

While the necessity of sanitation is urgent, the value of human excreta should not be disregarded, as excreta contains nutrients for agriculture such as nitrogen, phosphorus and potassium. Phosphorus in particular is a limiting resource.

Agricultural use of chemical fertilizer accounts for 80% of total consumption of phosphate rock in the world. The depletion of current economically exploitable reserves of phosphorus can be estimated at somewhere from 60 to 130 years (Steen, 1998). We need to recover resources from human excreta as much as possible.

The purpose of this study is to examine a new system for advanced sanitation, which can combine treatment of human excreta and recovery resources from it and which developing countries can afford. Japan has a sophisticated system to handle human excreta without a pipe system, although little attention has been given to the system internationally. The system has been installed in peri-urban and rural area in Japan, where sewage systems cannot be installed efficiently. We would like to re-evaluate this Japanese night-soil treatment system.

Interest in ecological sanitation (Winblad 1998) has been growing. Interest is directed to both treatment of human excreta and recovery of resources from it efficiently. It has been proposed in the concept of ecological sanitation that urine, the contents of which differ significantly from that of faeces should be diverted from faeces. We also examine faeces-urine separation in this study.

Based on the examination, we will propose a new system to treat human excreta in peri-urban areas of water-limited developing countries.

Night-soil treatment plant and vacuum truck collection system in Japan

Priority of treating human excreta to grey water

We will begin our discussion by considering the human excreta in household wastewater. Table 1 (Japan sewage works association, 1999) shows statistical data on the pollution load of excreta and grey water in Japan. As Table 1 shows, human excreta are main source of nitrogen and phosphorus while the others contain much of the BOD and COD. Nitrogen and phosphorus in wastewater should be treated properly since they induce eutrophication, and are targets for resource recovery, especially phosphorus. High concentration of ammonia (much of the T-N) requires nitrification and denitrification in the sewage treatment process. In addition, pathogens, which are contained in excreta, should be sanitized completely. Therefore, excreta are more important targets for treatment and effective management than grey water.

Due to the low concentration of ammonia, grey water could be treated by simple treatment system such as artificial wetland and simple treatment equipment with open channels. The priority problem in the treatment of household wastewater in peri-urban and rural areas in developing country is how to treat human excreta.

Table 1 Pollution loads of household wastewater (g/person/day)

	AVG.	S.D.	N	Excreta		Grey water	
BOD	58	18	125	18	(31.0%)	40	(69.0%)
COD	27	9	120	10	(37.0%)	17	(63.0%)
SS	45	17	125	20	(44.4%)	25	(55.6%)
T-N	11	2	19	9	(81.8%)	2	(18.2%)
T-P	1.3	0.3	15	0.9	(69.2%)	0.4	(30.8%)

Non-pipe system to treat human excreta

Sewage systems are a vital component of the disposal and treatment of household wastewater, although most areas in developing countries cannot immediately afford its installation. As Table 2 (Japan sewage works association, 2000) shows, the high cost of the system is derived from construction of the pipe system. This cost is borne by those areas which tried to install the sewage system. In addition, transport of excreta with the pipe system needs much water and it is, therefore, not suitable to water-limited area of developing countries.

Therefore, if a non-pipe system can perform properly, the non-pipe system can become a viable alternative to traditional piped sewage systems. In fact, Japan has a sophisticated non-pipe system to treat human excreta with vacuum truck collection from pit toilets, which has dramatically contributed to improve sanitary conditions in Japan. Now we will examine the Japanese system to treat human excreta.

Table 2 Total construction cost of sewage treatment plants and pipe systems in Japan

Year	Treatment plant Cost (billion yen)	(%)	Pipe system Cost (billion yen)	(%)
1996	980	27.2	2,618	72.8
1997	898	26.2	2,532	73.8
1998	986	27.2	2,638	72.8
1999	999	26.3	2,804	73.7
2000	860	25.2	2,548	74.8
AVG.		26.4		73.6

History of night-soil treatment and vacuum truck collection system in Japan

Japan has a history of using human excreta as fertilizer (Kaneko et al., 1998). In the 17th century at the latest, people established a sophisticated system to recover human excreta. Human excreta were handled separately; urine and mixed urine and faeces were valuables. Human excreta were transported from urban to rural areas and used as fertilizer.

However, the amount of human excreta used as fertilizer reduced due to the urbanization and the prevalence of chemical fertilizer. Excess human excreta were not treated appropriately and contributed to the deterioration of sanitary conditions especially in urban areas. As it was necessary to treat human excreta appropriately, and piped sewerage systems were not widely developed in Japan at that time, the Japanese government started to develop night-soil treatment plants in 1954 (Inoue, 2004). As people at the time were using pit toilets, vacuum trucks were used to transport contained human excreta to the plant, instead of pipe-system. From the 1950s to the 1970s, this system using vacuum trucks spread across Japan. Together with the development of water supply systems and improved nutrient conditions, the night-soil system dramatically contributed to a reduction in the rate of water-born diseases infection derived from human excreta by 1970s (Magara, 2003).

Thus, this unique Japanese system was build up. However, water-flushed toilets were introduced with the modernization after the 1970s by the development not only of sewerage systems but also of *johkaso.* Johkaso is an on-site treatment system and treats human excreta and/or grey water by anaerobic and aerobic biological system. However, sludge accumulated in *johkaso* tanks need to be transported to night-soil treatment plants by vacuum trucks and treated in the plants. The public sewage system has developed in urban areas and the other systems covered peri-urban and rural areas. In 2000, the excreta-disposal systems consist of public sewage (56.2%), *Johkaso* (27.2%), and non-flashed pit toilet (16.6%) (Ministry of Environment, 2003). Therefore, 43.8% of total population take a service of night-soil treatment system.

Existing night-soil treatment system in Japan

Vacuum truck collection

Pictures of the vacuum-truck collection service are shown in Figure 1. A vacuum truck is shown in Figure 1 (a). There are a range of trucks in the capacity range from 2 -10 tonnes and all trucks have deodorizing equipments to meet to requirement of odour control. In general, two people operate each truck.

A vacuum truck goes to each household once a few months to collect human excreta in each toilet pit. A truck stops in front of the each household and collects human excreta with waste paper for anal cleaning from the toilet pit, as shown in Figure 1 (b). The opening to remove night soil shown in Figure 1 (b) is placed at just next to the each household. After a truck collects excreta from some households, excreta are carried into a night-soil treatment plant. Sludge in *Johkaso* tanks is also collected in the same way.

(a) (b)

Figure 1 Pictures of vacuum-truck collection: (a) a vacuum truck in a night-soil treatment plant, (b) vacuuming human-excreta from a toilet pit.

Night-soil treatment plant

At the night-soil treatment plant, collected human excreta and *johkaso* sludge are treated. A flow diagram of the major night-soil treatment plant process is shown in **Figure 2** as an example of the latest advanced treatment. Human excreta and *johkaso* sludge are treated without dilution. In a nitrification-denitrification tank, human excreta are intermittently charged into the tank, and aerobic (for nitrification) or anaerobic (for denitrification) conditions are created in the tank intermittently. The residual NO_x-N from nitrifying tank is denitrified in the secondary denitrification tank with the use of methanol as electron donor. The re-aeration tank treats the residual BOD and residual methanol.

In order to keep high concentration of MLSS (12000 – 20000 mg/L) in the reaction tank and to treat wastewater in high concentration, the concentration of return sludge is increased by the use of ultra filtration or micro filtration in solid-liquid separation equipment (Japan waste management association, 2001). This process has two membrane filtration processes; one for biological treatment and for coagulation. The coagulation process removes phosphorus by addition of flocculants. This process also includes an activated carbon absorption chamber for the removal of COD as well as colour.

Scenes from a plant are shown in Figure 3. This plant has two lines of biological treatment process and total capacity is 168 kL/day. As shown in Figure 3 (a), recent night-soil treatment plants are indoor plants to minimize the impact of the plant's presence on the surrounding communities. The treated water quality of the plant is shown in Table 3. The quality is same or better than sewage treated water

Recently, the sludge treatment process has changed to a resource recovery process such as a composting process, methane fermentation process and carbonization process. These processes can receive garbage. The wastewater generated in the garbage disposal process can be treated at night-soil treatment plants.

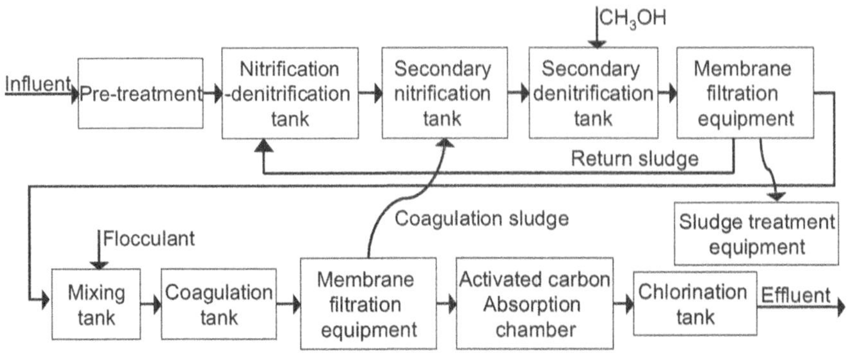

Figure 2 The process of a high-load night soil treatment by membrane filtration

Figure 3 A night-soil treatment plant using high-loading denitrification process: (a) Outside of a night-soil treatment plant, (b) a circulation pipe of a nitrification-denitrification tank, (c) coagulation equipments, (d) activated carbon chambers

Table 3 Quality of treated water of the plant

	Measured value	Standerd value
pH	7.3	6.0 - 8.5
BOD (mg/L)	1.2	20
COD (mg/L)	4.9	20
SS (mg/L)	3.9	70
T-N (mg/L)	6.3	10
T-P (mg/L)	0.4	1
Coliform group count (N/mg)	2	3000

Effectiveness of urine–faeces separation

Although pathogens found in human excreta bring about varieties of water-born diseases, their nutrient value should not be disregarded. Nitrogen, phosphorus and potassium are essential nutrients for agricultural activities and high concentration of them occur in human excreta. Therefore, combined treatment and sanitization of human excreta, and the recovery of nutrients from them are crucial to treat human excreta properly.

Table 4 (Maksimovic and Tejada-Guibert ed., 2003) shows the relative ratio of the principle nutrients - or nitrogen, phosphorus and potassium - of urine and faeces. The important fact is that urine contains higher amounts of essential nutrients even though most people imagine that there should be more nutrients in human faeces. In contrast, faeces contain a large amount of pathogens, although urine dose not generally contain pathogens that can be transmitted though the environment (Hoglund, 2001).

When urine is collected without mixing with faeces, it can be used as fertilizer easily with little risk regarding sanitation. In addition, it has been noticed that urine-faeces separation leads to efficient treatment and nutrient recovery (Matsui, 2001; Hellstrom, 1999). In addition, the amount of water used for flushing toilets is also reduced. Toilets in average use 6 - 8 litres of water for each flush, even for only urine, although the toilets based on separation system use only 0.2 - 0.5 litres for flushing urine. The systems can save water in water-limited environment. Thus, urine-faeces separation has possibilities for alternative to existing sewage system to treat human excreta.

Table 4 The ratio of nutrients between urine and faeces

	Urine		Faeces	
	g/person/day	%	g/person/day	%
Nitrogen	11.0	88	1.5	12
Phosphorus	1.0	67	0.5	33
Potassium	2.5	71	1.0	29

Combination of vacuum truck collection system and urine–faeces separation

The system to handle human excreta with vacuum-truck collection can be an alternative to sewage system in peri-urban areas, where sewage system is not affordable. Furthermore, urine-faeces separation can contribute the efficient treatment of wastewater and the recovery of resources from human excreta more efficiently than the conventional means, in which urine and faeces are mixed. We will now combine vacuum truck collection system and urine-faeces separation, and propose a new system to treat human excreta properly and recover nutrient from them.

A diagram of the new system is shown in Figure 4. Grey water could be treated by the simple treatment system such as artificial wetland with existing channels. The new system is suitable to areas where infrastructure such as car transportation, tap water and local electricity are available but sewage systems are unaffordable. Most peri-urban areas of developing countries meet such conditions.

Collected faeces could be treated in some ways. Japanese night-soil treatment can be one of the treatment means by which sanitation can be improved and water pollution can be prevented. In addition, a more efficient means of resource recovery can be proposed as shown in **Figure 5**. In the flow in **Figure 5**, separated faeces are treated by the fermentation of methane. By the separation of faeces from urine, the fermentation progresses more easily due to the low concentration of ammonia that would otherwise inhibit fermentation. In addition, garbage could be treated together with separated faeces. Recovered methane gas could be used for electricity to run this plant. Fermented sludge and liquid could be used for agriculture. The sludge may be further composted. All products and effluent do not contain much salt because they are separated from urine that contains much salt.

On the other hand, separated urine could be used with dilution to farmland directly. In addition, as shown in **Figure 6**, nitrogen and phosphorus could be recovered from it as powder of magnesium ammonium phosphate (MAP, or struvite, which is a slow-release fertilizer) by adding $MgCl_2$ and controlling pH in alkaline conditions (Harada in press; Sato, 2004) and liquid fertilizer ($(NH_4)_2SO_4aq$) by aeration in H_2SO_4aq. Concentrating nutrients of urine as fertilizer is effective as it facilitates its transportation and use in other areas. The residue liquid contains salt in high concentrations and could be disposed of at site of salt accumulation after evaporation of liquid.

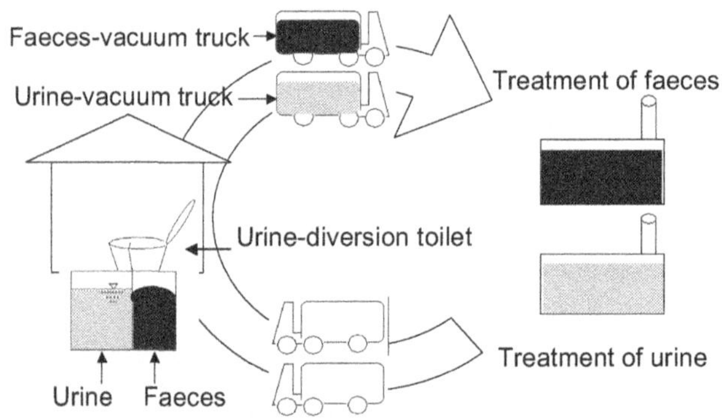

Figure 4 A vacuum truck collection system with urine-faeces separation

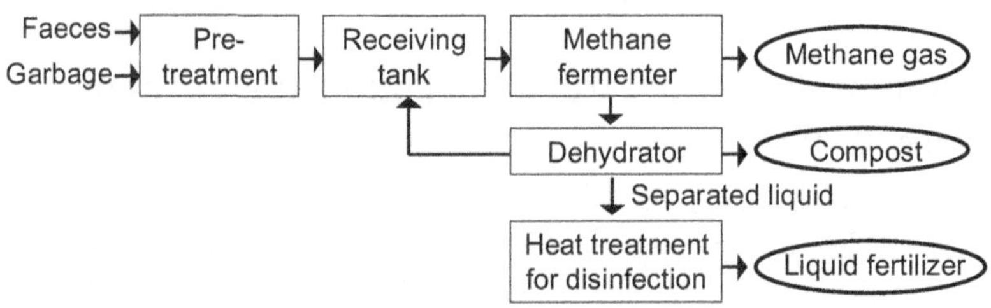

Figure 5 A treatment flow of separated faeces and garbage to recover resources

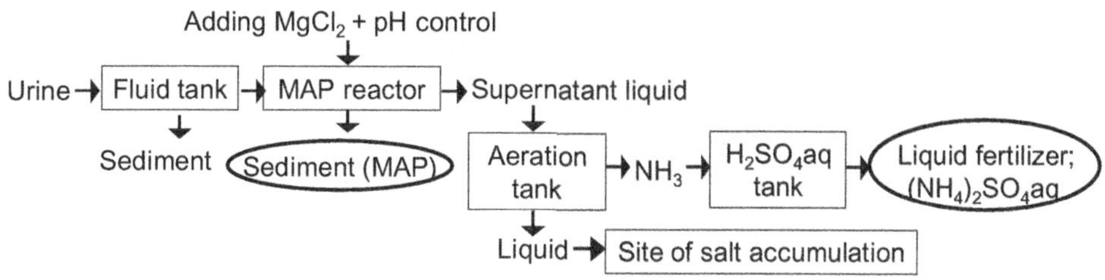

Figure 6 A treatment flow of separated urine to recover resources

Conclusion

In this study, we proposed advanced sanitation with the vacuum truck collection of human excreta, which can treat human excreta properly and recover resources from them. The system is suitable to peri-urban areas of developing countries. Effectiveness of the systems compared to the existing sewage system is summarized below:

- Installation cost is not as expensive as conventional sewage systems due to the use of vacuum truck collection instead of a piped collection system.
- This system can save water due to the use of vacuum truck collection instead of the pipe collection system.
- This system can treat human excreta in high efficiency due to urine-faeces separation.
- This system can recover resources from human excreta at high efficiency due to urine-faeces separation. Valuable products such as MAP, ammonia sulphate and methane are generated.
- Salts concentration in the produced effluent from the plant is low due to urine-faeces separation. The residue liquid is disposed of at salt accumulation site.

This paper drew a blueprint of advanced sanitation alternative to existing sewage systems. Further research on each treatment and system would clarify the possibility of vacuum-truck collection system with urine-faeces separation. This new system could contribute to find a solution to treatment of human excreta and recovery resources from them in peri-urban areas of developing countries, which may help to meet to MDGs.

References

Harada, H., Shimizu, Y., Miyakoshi, Y., Matsui, S., Matsuda, T. and Nagasaka, T. (2005) Prediction of MAP formation to recover phosphorus from human urine using an equilibrium model. *Journal of Japan Society on Water Environment*, **28** (3), 191–196.

Hellstrom, D. (1999) A comparison of source separation systems and conventional treatment systems, *Water Environment Research*, **71**(7), 1354–1363.

Hoglund, C. (2001) *Evaluation of microbial health risks associated with the reuse of source-separated human urine*, KTH and SMI, Stockholm.

Inoue, Y. (2004) History of night soil treatment plant and sludge recycling center, *Journal of Japan Waste Management Association*, **57** (261), 431–437.

Japan sewage works association (2000) *Gesuidou Toukei Youran* **57** (3), JSWA, Tokyo.

Japan sewage works association (1999) *Ryuikibetsugesuidouseibisougoukeikakuchousa shishin to kaisetsu*, JSWA, Tokyo.

Japan waste management association (2001) *Odeisaiseisyorisentatou sisetuseibi no keikaku·sekkei youryou*, JWMA, Tokyo.

Japanese Ministry of Environment (2003): *Ippanhaikibutu no haisyutu oyobi syorijyoukyoutou ni tsuite*, Tokyo.

Kaneko, M., Kawamura, K. and Nakajima, J. (1998) *Seikatsuhaisuisyorishisutemu*, Gihoudou, Tokyo.

Magara, Y. (2003) Status of onsite-treatment of domestic wastewater management in Japan. In *Proc. of the Johkasou session, the 3rd World Water Forum*, Kyoto, 17–30 March.

Maksimovic, C. and Tejada-Guibert, J. A. ed. (2003) *Frontiers in Urban Water Management Deadlock or Hope*, trans. Saburo Matsui et al., Gihodo shuppann, 231.

Matsui, S. (2001) A paradigm shift of urban waste water management for conserving freshwater resource against eutrophication. In *Proc. of the 2nd Civil Engineering Conference in the Asia Region*, Tokyo, 207–210.

OECD (2002) *OECD environmental data compendium 2002.*

Rauch, W., Brockmann, D., Peters, I.., Larsen, T.A. and Gujer, W. (2003) Combining urine separation with waste design: an analysis using a stochastic model for urine production. *Water Research*, **37**, 681–689.

Sato, D. (2004) *Removal and Recovery of Nutrients from Human Urine and Pig Waste.* MSc thesis, Kyoto University.

Steen, I. (1998) Phosphorus availability in the 21st century Management of a non-renewable resource, *Phosphorus & Potassium*, no. 217.

Winblad, U. and Simpson-Hebert, M. (ed.) (2004) *Ecological sanitation- revised and enlarged edition.* Stockholm Environmental Institute, Stockholm.

A sustainable approach towards rural development: dry toilets in Nepal

Regmi Megha Raj

GPO BOX 4836, Bagmati, Kathmandu 36987, Nepal. (E-mail: megha_rajregmi@yahoo.com)

Abstract Existing inadequate sewerage systems and direct disposal of household waste into the watercourses has significantly increased water pollution. Dry toilets are feasible in the rural and peri-urban areas of Nepal to reduce the consumption of costlier water that is required for flushing. As conventional treatment technologies require high investment and operation and maintenance costs, the inexpensive dry toilets are the only suitable option left for sanitation for the 21st century. To reduce environmental degradation and reduce costs dry toilets are the only the realistic option in the context of Nepal.

Two reactors, one exposed to sunrays and the other without sunrays were constructed. In the sunrays model, it was found that in 48 days of observation the faecal coliform presence depleted to 610 cells per gm from the initial value of 7×10^{10} and the volatile organic matter came down to 70.18% from 98.09%. Similarly, in the other model the destruction of faecal coliform in 65 days (and was found to be 920 cells/g) while the destruction of organic matter took 75 days. Also from the observation of 313 people who formed of a cluster of the pilot project, annually recovered value of N, P and K was found to be 1565 kg, 125 kg and 344 kg, respectively.

This paper deals with the different types of Dry ecological toilets, their performances and feasibility study in the context of Nepal, with full involvement of local people based on complete laboratory analysis and regular monitoring. Using Dry toilets will save 14 lpcd, which is equivalent to 14 MLD or 80 MLD for the urban area of Kathmandu only. The result supports the implementation of ecological dry toilets to save the valuable water wasted in flushing as well as the resources used to treat the waste.

Keywords Dry toilet; faeces biodegradation; human excreta reuse; pathogen destruction

Introduction

As the same time as a significant number of people in the world are still struggling for a bucket of water, most urban populations are using precious water to flush excreta and urine. Using flush water means converting potable water into more waste and if the waste is left inappropriately treated, making rivers and water bodies into open sewers. Socially, it is a crime to poor and deprived people. To overcome this problem, new skills are required for improved sanitation services, as conventional wastewater treatment technologies require high investment and proper operation and maintenance cost.

In Nepal, the water supply situation of the whole country is not that bad if compared to bleak situation of sanitation. Since the government has invested substantially in water supply systems with a target of 100% supply by the year 2002, water access in rural area has been significantly improved. However, the service delivery is not up to level in urban area because urbanization in occurring more rapidly than water supply development. The urbanization rate of Nepal is 6.5% annually which is the highest figure among the SAARC countries.

It is obvious that if we have actual demand, one can have a clear idea of efficient management. Thinking that water is the most precious but scare natural resource, the distribution, use and reuse of water should be properly guided. To reduce the environmental degradation of the rivers and others water bodies, to conserve the water sources and to manage the available water, dry toilet is one of the best viable options.

Demand for water in Kathmandu valley is 200 million litres per day (MLD), whereas the supply is about 120 MLD and 15% of supplied water is used to flush the excreta. In the Kathmandu valley all the rivers including the Holy Bagmati River now look like an open sewer due to the direct disposal of untreated sewage into the watercourse. To reduce the environmental degradation dry toilet is the only realistic option in the context of Nepal. Dry toilet is defined in this report as the on-site disposal of human urine and faeces without the use of water as a carrier. The dry toilet not only lessens the pollution in the water course but also enables the reuse of the urine & faeces as nutrient & conditioner to increase the fertility of soil. This could eventually reduce the haphazard use of chemical fertilisers.

Methodology

The water demand of the valley is different in different literature, but for the purpose of this paper, the most recent survey, literature review, interviews with key personnel and informal discussion were taken into consideration. The entire process of the study of the dry toilets was carried out in a set of theoretical and experimental research works.

Future water demand of Kathmandu

Water consumed by the households in the 24 h supply area could be taken as the trend in potable water use that people would follow when there is adequacy in supply with good quality water. However, the probable rise in water price in future will be again a repellent in increase in water consumption. Economic improvements will certainly raise the quantity of water consumed. For an average figure the domestic water consumption may be considered 94 lpcd for the future projects, which however need to take into account of the trend of development of this capital city.

Dry sanitation

Dry toilet research has been conducted in the peri-urban area of Kathmandu Valley. Beneficiaries were 313 people in a cluster. Ten toilets have been constructed. The study of the performance of toilets has been found successful. People have been using source separated urine and dehydrated faeces in the agricultural field. The study shows that a high nutrient value of the excreta and farmers are currently using it as fertilizer in place of chemical fertilizer in the vegetables and crops. The temperature and seasonal variation of the Kathmandu Valley is very much conducive to running a dry toilet. We are planning, 50 such toilets to be constructed in this year in this area. The study confirms that the dry toilets have saved the water for flushing at the minimum rate of 14 lpcd in this local context for the peri-urban of Kathmandu valley.

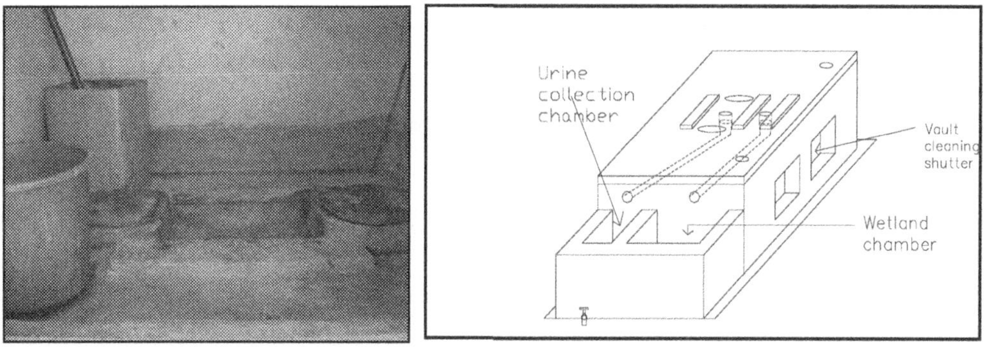

Figure 1 Isometric view of dry toilet

Laboratory analysis

In the laboratory analysis methodologies adopted for the study are as follows:

(i) Faecal coliform of urine and faeces analysis were carried out using the multiple tube fermentation technique.

(ii) The analysis of pH of urine and faeces sample, Electric Conductivity and total dissolved Solid of urine sample was conducted. Similarly the biodegradation of faeces was examined by ignition, using Muffle oven at 550 °C.

(iii) Total Nitrogen, Phosphorus and Potassium, TOC were also measured.

Performance evaluation

In order to evaluate the function of constructed dry toilets, questionnaires were prepared and distributed to the concerned household and analysed. The responses obtained from the survey were later used to rectify the shortcomings in the design.

Results and discussion

Urine analysis

The analysis of Nutrients (NPK) showed that total Nitrogen, Phosphorous, and Potassium were found up to 14.21 g/L, 0.713 g/L and 2.51 g/L, respectively. The results indicate that the sufficient concentrations of major plant nutrients (NPK) are present in urine. In case of flushing toilets these nutrients are wasted and become of a cause of contamination. So, the above result proves the dry toilet to be an ecologically better option.

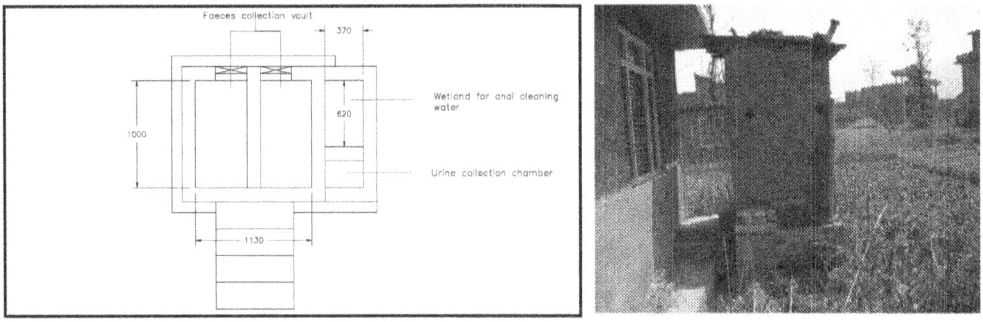

Figure 2 Plan of implemented dry toilet

Faeces analysis

Biodegradability of faeces was carried out taking the sample from the small-scale reactor. During the lab test sample was weighed and taken to the muffle oven at 550 °C for two hours. Faeces of 25 days storage contain organic matter 93.74%, after 47 days the percentage of Volatile organic matter was found 70.18%. Similarly in reactor-2 36 days storage faeces contains 93% volatile organic matter. Whenever after 75 days storage contains 70.56% volatile organic matter. Total organic carbon content of the first day faeces experimental results shows 56.89% and 48 days faeces 40.71%. Obtained results of reactor one is presented below in the Figure 4 and reactor two in Figure 5:

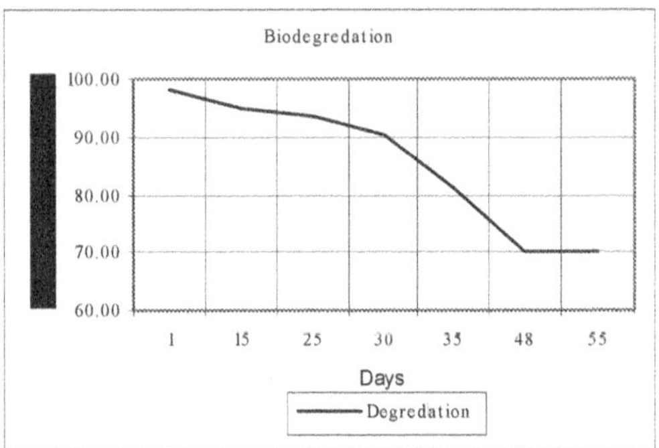

Figure 4 Biodegradation of volatile organic matter reactor-1

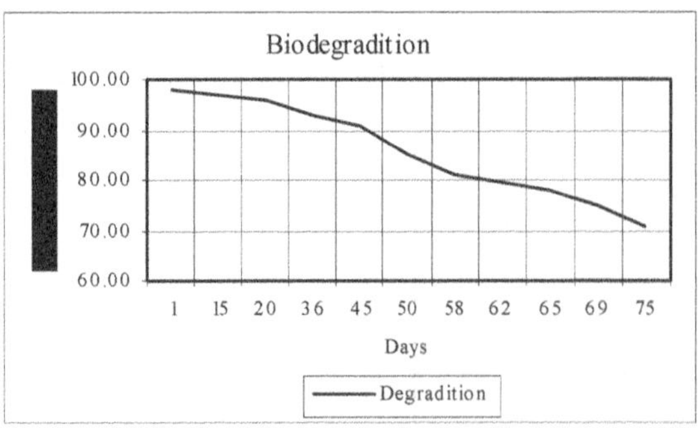

Figure 5 Biodegradation of volatile organic matter reactor-2

Temperature variation analysis

To analyse the temperature variation of faeces, monitoring of temperature of urine-separated faeces was conducted on both the reactor constructed with and without sunrays. The results of the temperature suggest that when the ambient temperature is 22 °C the inside temperature of the reactor rose up to 40 °C. This fact reinforced the concept of dry toilets creating a speedy decomposition.

Faecal coliform analysis

In reactor one, to analyse the pathogen destruction, sample of urine-separated faeces was taken from the reactor with sunlight and laboratory test was conducted by multiple tube fermentation technique. The result of the faecal coliform concentration on first day was 7.0×10^{10} cells/g, after 35 days the concentration was found to be 1.7×10^{7} cells/g and 6.1×10^{2} cells/g after 48 days. Similarly in reactor two, the result of the faecal coliform concentration on the first day was 3.5×10^{13} cells/g, after 35 days storage the concentration was found to be 1.7×10^{12} cells/g and 3.6×10^{2} cells/g after 70 days. The detail is presented in Figures 6 and 7.

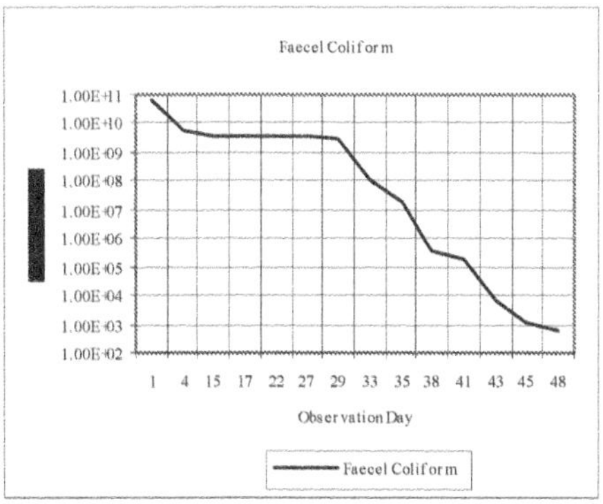

Figure 6 Pathogen destruction in urine-separated faeces with sunrays

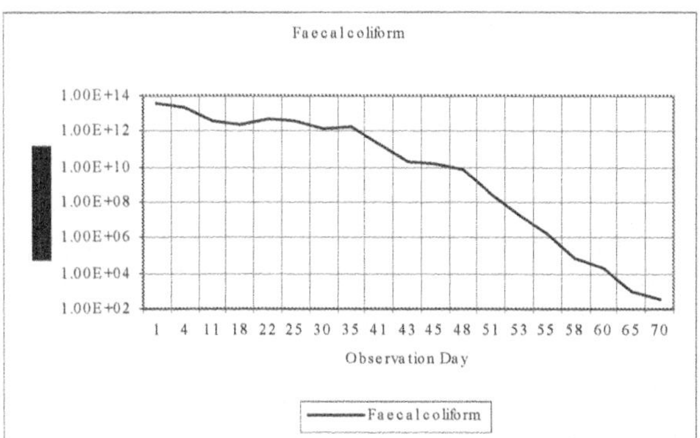

Figure 7 faecal Coliform reductions in urine-separated faeces without sunrays

Feasibility of the dry toilets

The users' perception regarding its ease in uses and their acceptance in comparison to traditional type of pit system toilet clearly revealed the feasibility of the Dry Toilet concept. Since this type of system has been introduced for the first time in Nepal, it has been necessary to get the peoples' response regarding its ease in construction and uses, odour and foul problem and the potential of recovery of the nutrient value from faeces and urine and use it in agricultural crop farming. The result of the base line survey has been used to arrive at the feasibility of the Dry Toilets.

Public Response to the dry toilets

Public response study was carried out by sample household survey of the users of dry toilet in the pilot area. The questionnaire particularly focussed on the acceptance of the toilet by the users. Out of total 65% of the respondents were in the view that the dry toilet is better than the traditional type. However, 25% suggested design modification. The modifications suggested by the respondents were towards the size of toilet and shape of the pan hole.

Modified Plan and Section of Dry Toilets

As per the people's recommendation the modified plan and section comes out to be as:

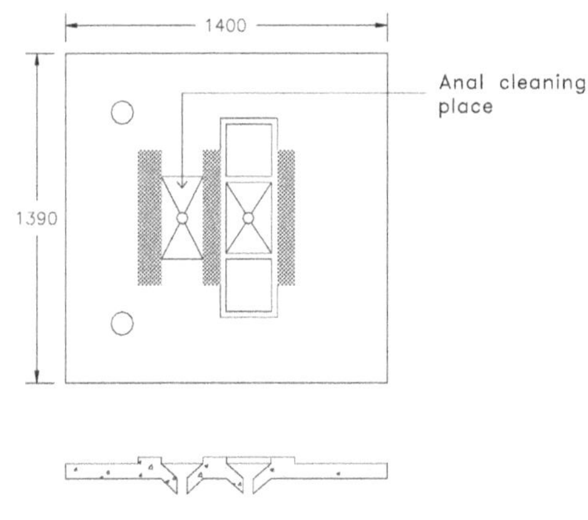

Potentials of Resource Recovery

The objective of this research work was to develop a mechanism so that the nutrient value of the urine and faeces is recovered so as to use it in place of chemical fertiliser. On average an adult produces about 500 litres of urine in a year. On the basis of the lab test 500 litres of urine contains 5.0 kg of Nitrogen, 0.4 kg of phosphorus and 1.1 kg of potassium. Total population according to survey of the cluster area is 313. This means 1565 kg of nitrogen is produced annually. Similarly, phosphorus and potassium production annually amounts to 125 kg of 344 kg, respectively. This can be expected to suffice the local fertiliser needs. The above discussion clearly shows that the Dry Toilet system can have high potential of Nutrient recovery and can reduce the use of chemical fertiliser some extent.

Cost effectiveness

The comparative cost analysis of different toilets was carried out. The result of the analysis shows that the dry toilet is cheaper compared with all the other types of toilet except offset type VIP pit latrine. The comparison is such that the next cheaper option, latrine with small septic tank (€112), is 35.5% costlier than the dry toilet (€83).

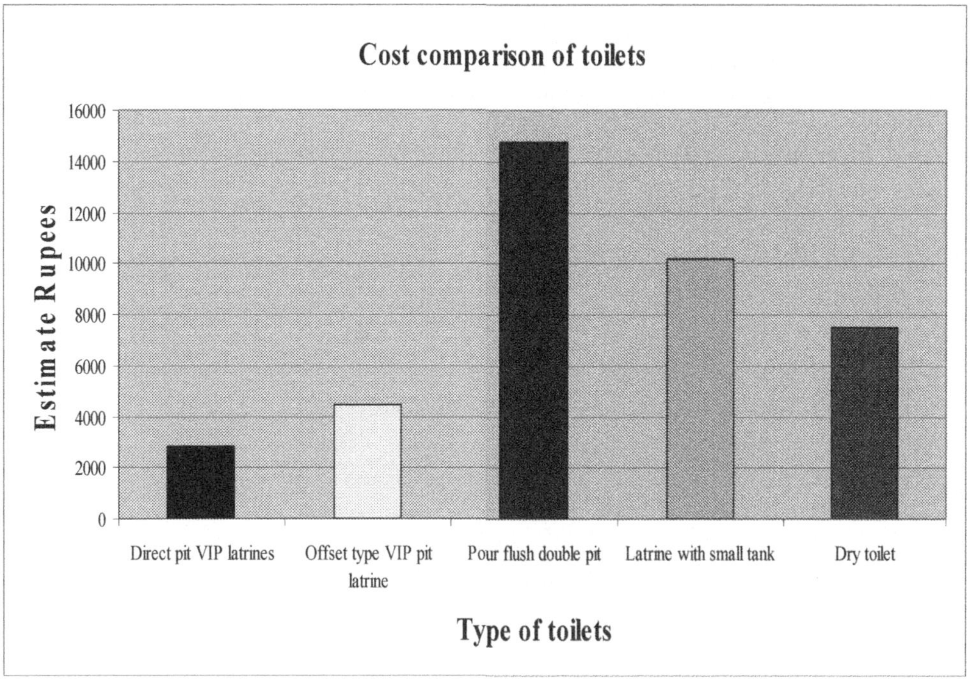

Conclusions

The conclusions drawn from the study is as follows:

(i) The actual water demand found by the recent survey is about 94 lpcd.

(ii) The decomposition process of faeces start very rapidly at the temperature above 22 °C.

(iii) The decomposition rate was 1.5 times faster in the vault with sunrays toilet than without sunrays one. To arrive at 80% reduction of organic matter in faeces it took 48 days in reactor 1 compared to 70 days in reactor 2.

(iv) The destruction of 99.9999% faecal coliform was achieved within 48 days in the reactor 1 and 70 days in reactor 2. This indicates that reactor 1 is more efficient. A retention time of two months seemed short for total pathogen destruction in without sunrays toilet

(v) About 80% of water consumption can be reduced in the Dry Toilet compared with other toilets. This can tremendously benefit the areas with water scarcity.

References

Aussie, A. and Louiza, D. (2002) Urine Diversion ecological sanitation systems in South Africa. CSIR, Pretoria, South Africa.

Ecological sanitation (1998) A publication of Swedish International Development Cooperation Agency, Stockholm.

Joshi, P.S., Shrestha, K.B. and Shrestha, P.L. (2003) Household Water Use Survey and Research in Urban Kathmandu Valley, NGO Forum, Kathmandu.

Regmi, M.R. (2003) Exploration the concept of dry Sanitation, Kathmandu, Nepal.

Shrestha, M.S. (2002) The water supply problem in Kathmandu Valley. Kantipur Daily, Kathmandu, Nepal.

Separate discharge and treatment of urine, faeces and greywater: demonstration project in Berlin, Germany

A. Peter-Fröhlich*, I. Kraume*, F. Luck*, A. Lesouëf* and M. Oldenburg******

*Berliner Wasserbetriebe, Wastewater Works, Neue Jüdenstraße 1, D 10864 Berlin, Germany (E-mail: anton.peter-froehlich@bwb.de)
**KompetenzZentrum Wasser Berlin gGmbH, Cicerostraße 24, D 10709 Berlin, Germany (E-mail: francis.luck.@kompetenz-wasser.de)
***Veolia Water, WWTreatment Department, 18 Boulevard Malesherbes, F 75384 Paris, France (E-mail: andre.lesouef@generale-des-eaux.net)
****Otterwasser GmbH, Engelsgrube 81, D 23552 Lübeck, Germany (E-mail: info@otterwasser.de)

Abstract The Berlin Centre of Competence for Water (Kompetenzzentrum Wasser Berlin) together with its partners Berliner Wasserbetriebe and Veolia Water has started a demonstration project about new sanitation concepts. To define the experiments for testing new, sustainable sanitation concepts a pre-study has been performed. This study included a cost comparison between two new sanitation concepts with gravity and vacuum separation toilets and the conventional system. It could be demonstrated that the new sanitation concepts may have cost advantages depending on the situation. This was a further motivation to start a Demonstration project near Berlin testing the innovative toilet systems under realistic conditions. Operation of the gravity separation toilet concept started in October 2003.

Keywords Cost comparison; decentralised wastewater concepts; gravity separation toilet; urine; vacuum separation toilet

Introduction

Conventional centralised concepts for water supply and wastewater, developed for ages in industrialized countries, imply high costs and high water consumption, which make them not suitable as a sustainable solution especially for developing countries. Further development, testing and dissemination of alternatives to conventional wastewater systems are therefore becoming more and more indispensable for ecological, economic and societal reasons. More sustainable approaches should consider reuse of treated water as well as nutrient recycling if possible. Furthermore energy consumption for wastewater discharge and treatment should be minimised. Such techniques and concepts are already available and in use, but further developments and validations are necessary.

These are the reasons why the Berlin Centre of Competence for Water initiated with Berliner Wasserbetriebe and Veolia Water a demonstration project for new sanitation concepts which should represent a relevant solution for:

- remote areas, where the connection to a central system (e.g. large sewer networks) would not be technically or economically interesting,
- rapidly growing conurbations in developing countries,
- countries with scarce water resources and
- a contribution to the sustainable development with the recycling of nutrients and water.

Methods

The project is divided into two phases (***Phase I,*** pre-study and ***Phase II***, demonstration project). ***Phase I***, a theoretical approach, has been completed at the end of 2001.

- **A literature-based project review**, patent reviews and a collection of information on the various projects were made. Furthermore existing projects with separate treatment in Germany, Denmark and Sweden have been visited and the experience has been exchanged. These informations and detailed economic investigations have been the prerequisite for continuing the project.
- **Cost comparisons between conventional and new sanitation concepts** for an intended new housing estate in the federal state Brandenburg near Berlin have been made. The housing estate should be realised stepwise from 672 up to 5,000 inhabitants within the next 10 years. For the economic calculation three different sanitation concepts have been compared for different cost levels:
    - **Conventional sanitation concept:** Conventional flush-water toilets with flush-stop bottom, one sewer system, normal gravity sewer system for the area, pumping station with transport sewer to the existing sewer network, system operated by the public supplier.
    - **Separation sanitation concept (gravity, composting of faeces):** Gravity separation toilets with separate outlets for urine and faeces, collection and storage of the urine, transport to the farmer nearby and utilisation in agriculture, faeces transported by a gravity sewer system, aerobic treatment in a faeces separator, utilisation of the compost in horticulture on the area, transport of greywater in gravity sewer system, treatment in a constructed wetland, transport to the receiving water.
    - **Separation sanitation concept (vacuum, digestion of faeces):** Vacuum separation toilets, gravity urine transport, storage of the urine, transport to the farmer nearby and utilisation in agriculture, faeces transported by a vacuum sewer system, common treatment with ground bio-waste in a biogas plant, biogas utilisation by the equipment of the energy concept, transport of the digested sludge to the farmer nearby and utilisation in agriculture, transport of greywater in gravity sewer system, treatment in a constructed wetland, transport to the receiving water.

The cost comparison of these three sanitation concepts has been considered for four scenarios which are summarised in Table 1.

Table 1 Scenarios for the cost comparison of the three different sanitation concepts

	Inhabitants	**Water Operator ***
Scenario 1	672	Local company
Scenario 2	5,000	Local company
Scenario 3	672	Berliner Wasserbetriebe
Scenario 4	5,000	Berliner Wasserbetriebe

*This determines the costs of the "conventional" system and the costs for drinking water in all scenarios

A decision based on an economic point of view should consider three aspects:

- Costs of the investment
- Costs of the reinvestment
- Operation costs

These costs have been considered and the following assumptions were made for the cost calculations:

- Lifetime of the project: 50 years
- Duration of the components depending on their lifetime. Reinvestment after the end of the lifetime
- Real interest rate: 3.5 % per year
- Maintenance costs were calculated as a percentage rate of investment. Personal costs were taken in consideration separately.
- Operation costs divided into costs for
  - personal equipment
  - maintenance
  - water and wastewater
  - electricity
  - others equipment

The specific costs for water, wastewater, connection fees, energy and other costs were based on the information of the local company or of the Berliner Wasserbetriebe. With these assumptions and information the costs for the whole project period were calculated as dynamic prime costs (DPC). The DPC are the sum of money which is necessary for financing the whole project (investment, operation, reinvestment; see Figure 1) for the assumed lifetime based on today's cost level.

The calculations have been realised with the German guideline "Dynamische Kostenvergleichsrechnung" (dynamic cost comparison calculation) published by the "Länderarbeitskreis Wasser LAWA", a working group of all German federal states concerning water management) (LAWA, 1998). This method can also described as a life cycle cost analysis.

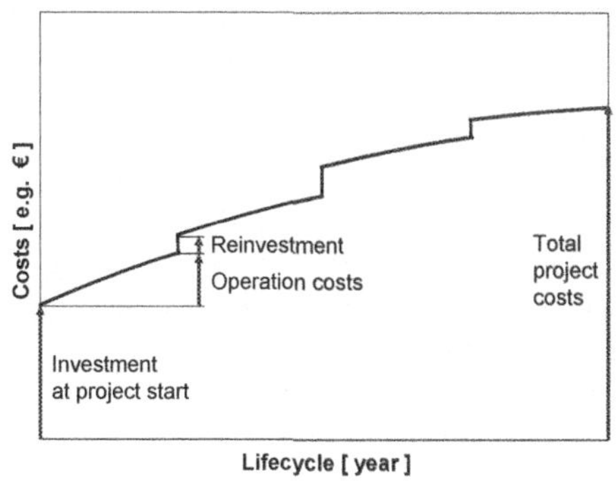

Figure 1 Demonstration of the total project costs with the DPC method

Results of the pre-study

Literature survey and visits

The main results from the pre-study are:

- 17 new sanitation projects already exist in Western Europe (Peter-Fröhlich *et al.*, 2003).
- Activities in relation to new sanitation concepts are increasing all over the world.
- Separation of urine, faeces and greywater based on the use of new toilet bowls, has proven to be feasible and accepted by the users (Johansson, 2001; Hellström and Thurdin, 1998; Swedenviro, 2001).
- Once urine is separated from the faeces, several configurations exist, differing by the collection and transport systems chosen and the treatment of the three effluents (faeces, urine, greywater) (Otterpohl *et al.*, 1999).

Cost comparison

The cost comparison between the conventional and new sanitation concepts shows that the new sanitation concepts have not only ecological advantages but can also have economical advantages (see below). The cost advantage is very depending on the specific conditions of the housing estate. For the chosen example (see methods) cost advantages occur in the most cases of the new sanitation concepts (Peter-Fröhlich *et al.*, 2003). This is show exemplarily in Figure 2 for *Scenario 2.* Both new sanitation concepts lead to cost savings compared to the conventional system after 3 and 9 years, respectively.

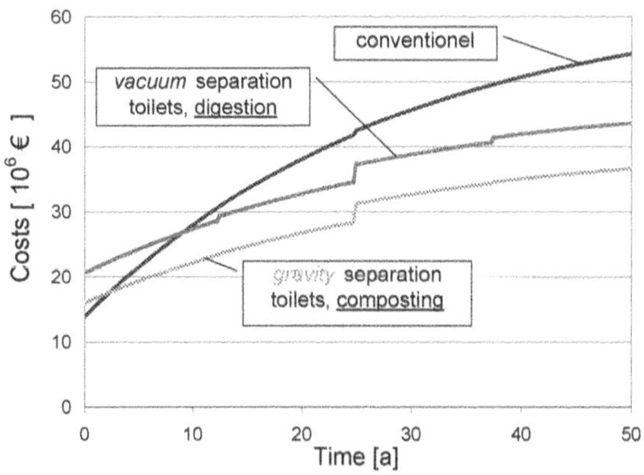

Figure 2 Total project costs for the conventional and the new sanitation concepts (*5,000* inhabitants); cost basis: *local company*

Demonstration project

The results from this cost comparison were an additional motivation to start a demonstration project with detailed evaluation of:

- gravity separation toilets and
- vacuum separation toilets

in conjunction with different treatment configurations.

The demonstration project started in the year 2002 (***Phase II***). The new sanitation concepts will be tested in existing buildings (office building and apartment house) of the Stahnsdorf WWTP owned and operated by the Berliner Wasserbetriebe. Realisation of the new sanitation concept in the office building took place during an extensive restoration of this building. In addition the apartment house will be partly equipped with the new sanitation concept until the end of 2004.

The general process scheme for the new sanitation concepts in the office building and in the apartment house can be seen in Figure 3.

In the new sanitation concept for the office building gravity separation toilets (10 toilets) are in use (Roediger, 2001). In addition, five waterless urinals from different suppliers have been installed in the mens toilets.

The faeces (brown water) are drained and composted. The filtrate from the faeces separator flows in the first chamber of the two-stage septic tank. It can also be filtered before mixing up with the greywater. Greywater is settled in a septic tank before treatment in a constructed wetland. In parallel to the constructed wetland a membrane bioreactor will also be tested for greywater treatment. Urine flows into storage tanks. Different methods will be tested for handling and treatment of urine before using it as fertiliser. The methods are vacuum evaporation, steam stripping, precipitation, ozonation, UV-irradiation and different combinations of these processes.

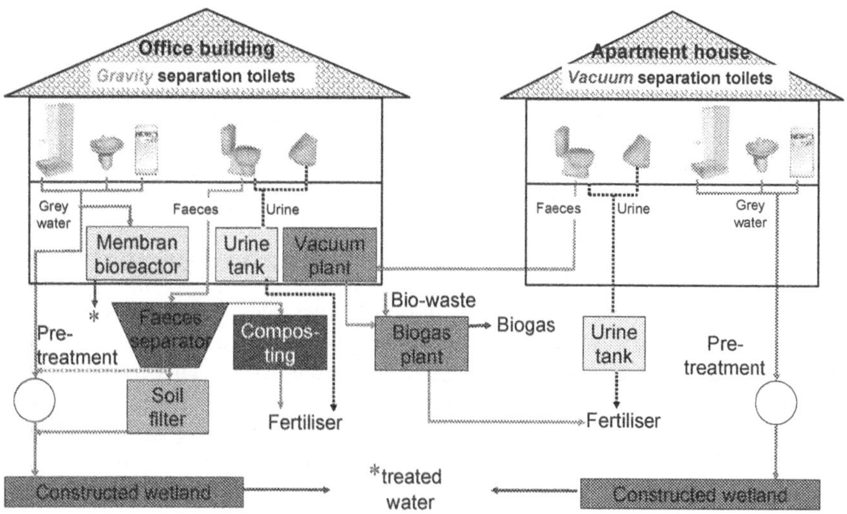

Figure 3 New sanitation concepts with *gravity* separation toilets in the office building and with *vacuum* separation toilets in the apartment house of the WWTP Stahnsdorf

For the new sanitation concept for the apartment house (15 flats) vacuum separation toilets were taken into consideration. In this concept urine and greywater are discharged and transported by gravity, while faeces are transported by a vacuum system. Urine will be treated as mentioned above. Faeces will be digested together with bio-waste. Digested sludge is also a fertiliser, e.g. for farmlands. Biogas can be used either in gas cookers or in a combined heat and power unit (CHPU). This topic will not be tested in this project. Greywater passes like in the case of the office building through a septic tank before treatment in a constructed wetland.

Since dish washing powders have a high content of phosphate (often more than 30 %) and dishwashing machines are more and more common, for both concepts a phosphate precipitation could also be necessary during greywater treatment.

The treated greywater can be used e.g. for irrigation in general. In this project the effluent of the membrane bioreactor will be investigated with respect to the different options of re-use as water with a lower quality than drinking water.

These two sanitation concepts are technical options belonging to the new approach, others are possible, e.g. composting of the faeces together with bio-waste should a production of biogas not be wished.

The type of greywater treatment for both new sanitation concepts will differ depending on the local situation. For large settlements an activated sludge tank etc. could be a more appropriate solution than a constructed wetland. The size of an activated sludge tank for greywater treatment could be however much smaller than for municipal wastewater treatment due to the much lower COD, nitrogen and suspended solids loads (Otterpohl, 2001).

As mentioned above vacuum separation toilets have been considered initially for the apartment house. Meanwhile, since vacuum separation toilets are still not available on the market, the concept has been changed: the longer experienced gravity separation toilets will be used instead. Until now only vacuum separation toilets are available from the company Roediger, which are modified gravity separation toilets. Furthermore, in the apartment house, only 10 flats instead of 15 will be used to test this sanitation concept.

The sanitation concept with vacuum separation toilets will now be tested in the office building. This is possible since a vacuum system is also installed in addition to the gravity system in this building. To change the concept only gravity separation toilets have to be replaced against vacuum separation toilets. The realisation of this is intended at the end of 2004.

Operation and first results of the demonstration project

Ten gravity separation toilets in the office building have been in operation since October 2003. For preliminary evaluation of a vacuum separation toilet, a gravity separation toilet has been replaced by a vacuum separation toilet in December 2003. This toilet is a modified gravity separation toilet from the company Roediger (see above). Urine is stored in tanks in the cellar of the office building. Faeces separator, soil filter, septic tank and constructed wetland have been in operation since the middle of March 2004. The soil filter could only be briefly operated since it was blocked due to high suspended solids concentration of about 250 mg/l in the filtrate of the faeces separator. The filtrate was not mixed up with the greywater but discharged to the influent of WWTP of Stahnsdorf until 4 September 2004. Since that time it is pumped into the first chamber of the two stage septic tank and treated together with greywater in the constructed wetland. The operation of the membrane bioreactor in parallel to the constructed wetland will start at the beginning on 2005.

The above-mentioned investigations for urine treatment have already been started by the Technical University Hamburg-Harburg. The Humboldt University Berlin began experiments on the efficiency of the different fertilisers in May 2004. Furthermore a life-cycle assessment was started at the Technical University Berlin.

The first results of the constructed wetland are shown in Table 2.

Table 2 Influent and effluent values (mean values) of the constructed wetland treating greywater from the office building (29.3.-26.7.2004)

Parameter	Unit	Influent	Effluent
Flow (disigned)	l/d	4,580	
Flow (actual)	l/d	1,321	
SS	mg/l	17.8	3.4
COD	mg/l	85	28
NH_4-N	mg/l	7.1	0.08
NO_2-N	mg/l	0.06	0.06
NO_3-N	mg/l	0.30	7.2
P-Total	mg/l	2.9	0.23
Total coliforms	CPU/100 ml	10^6	< 10
Salmonella (100 ml)		detection of Salmonella Gr. B and Gr. C	negative
Clostridium perfringens	CPU/100 ml	≤ 180	negative

Since the apartment house will not be connected before the end of 2004, the actual flow is below design flow. The chemical parameters show low values in the influent. These values may increase after connecting the flats due to greywater from washing machines and more greywater from the kitchens. The effluent values are in general very low. Nitrification was complete but no denitrification was observed. This is not surprising since the constructed wetland is under-loaded at present which prevents anoxic reactions. The very low concentration of P-Total in the effluent is typical for initial stages of the operation of constructed wetlands. Pathogen germs in the effluent remained below EU-guidelines for bathing water (EU, 1991).

Concerning the acceptance of the new sanitation facilities the users in the office building could fill a questionnaire sheet. Up to date most of them accept the gravity separation toilets in general. Improvements are necessary mainly in relation to flushing. Due to the fact that not all women sit on public toilets the opening system for the urine valve should also be changed, to allow urine collection. The experience with the vacuum separation toilet showed that flushing was not sufficient. After having been improved in May 2004 it is now acceptable but further improvements have to be done.

Conclusion

The results from the pre-study of this project enhanced the motivation for realising a pilot project testing gravity and vacuum separation toilets. Ten gravity separation toilets and a prototype of a vacuum separation toilet have been in operation since the end of 2003. The sanitation concept for the apartment house will be realisedbefore the end of 2004.

Although the cost comparison showed higher costs for the new vacuum sanitation concept compared to the gravity sanitation concept and in spite of a more complex operation, this system may be appropriate especially in case of local water shortages. Flush water consumption should be about 6 l/(p•d) compared to about 15 l/(p•d) for the gravity separation toilet or about 25 – 40 l/(p•d) for the conventional toilet with stop bottom.

Important objects of this pilot project (***Phase II***) are:

- Increasing knowledge on design, installation and costs of new sanitation techniques based on separation technologies
- Experience of the operation of new sanitation concepts by investigation of the various modules of the separation concept in different conditions.

The interest for these new sanitation concepts is increasing in general and further new projects are realised [e.g. Christ 2003].

Acknowledgements

This project is supported by the LIFE financial instrument of the European Commission (Duration: 1 January 2003 – 30 June 2006, LIFE03 ENV/D/000025).

References

Christ, O. (2003) Hochtechnologie für Bürohaus – Stoffsztomorientierte Abwasserbehandlung. wwt 10-11, 26-32.

EU (1991) Badegewässerrichtlinie - Richtlinie 76/160/EWG des Rates über die Qualität der Badegewässer vom 8. Dezember 1975 zuletzt geändert am 23. Dezember 1991 durch Artikel 3 der Richtlinie 91/692/EWG des Rates zur Vereinheitlichung und zweckmäßigen Gestaltung der Berichte über die Durchführung bestimmter Umweltschutzrichtlinien.

Hellström, D. and Thurdin, J. (1998) Drying as a urine dewatering method (Torkning som metod för att avvattna urin). Vatten, Vol. 54, 109-114.

Johansson, M. (2001) Urine Separation – Closing the Nutrient Cycle. Final report on the R&D project Source-Separated Human Urine – a future source of fertiliser for agriculture in the Stockholm region? Stockholm Vatten, AB Stockholmshem and HSB National Federation.

LAWA (1998) Leitlinien zur Durchführung dynamischer Kostenvergleichsrechnungen. Länderarbeitsgemeinschaft Wasser, Kulturbuchverlag Berlin GmbH.

Otterpohl, R., Abold, A. and Oldenburg, M. (1999) Source Control in Urban Sanitation and Waste Management: Ten Systems with Reuse of Resources. Wat. Sci. Tech., Vol. 39, No 5, 153-160.

Otterpohl, R. (2001) Black, brown, yellow, grey – the new colours of sanitation. Magazine of the International Water Association, WATER 21, October.

Peter-Fröhlich, A., Kraume, I., Lesouëf, A., Gommery, L., Phan, L. and Oldenburg, M. (2003) Sanitation Concepts for Separate Treatment of Urine, Faeces and Greywater. 2nd International Symposium on ecological sanitation in Lübeck, April 7 to 11, www.gtz.de.

Roediger (2001) Roediger-No Mix Toilet. www.roevac.com.

Swedenviro (2001) Market survey – extremely low flush toilets. Swedenviro report, No. 2001:1, February.

Udert, K., Larsen, T. and Gujer, W. (2002) Biologically Induced Precipitation in Urine-collecting Systems and Urinal Traps. IWA conference in Melbourne, Preprints

Use of constructed wetlands in treating recirculating aquaculture water for in-door intensive shrimp production

Ying-Feng Lin*, Shuh-Ren Jing*, Der-Yuan Lee, Yih-Feng Chang** and Wei-Chih Chen***

*Department of Environmental Engineering and Science, Chia-Nan University of Pharmacy and Science, Tainan 717, Chinese Taiwan
**Department of Environmental Resource Management, Chia-Nan University of Pharmacy and Science, Tainan 717, Chinese Taiwan

Abstract A commercial-scale recirculating aquaculture system was installed to demonstrate the effectiveness of constructed wetlands in treating the recirculating aquaculture water and the success in production of Pacific white shrimp (*Litopenaeus vannamei*). The system consisted of an indoor culture tank (64 m^2) and an outdoor wetland based water treatment unit that included a settling cell (1.5 m^2), a free water surface (FWS) wetland (12 m^2), a subsurface flow (SF) wetland (15.7 m^2), and a sump (1 m^2). Throughout the whole study of 126 days, the shrimp feed totalled to 328 kg and final shrimp body weight reached a food size of 12.3 g in average. Hydraulic loading rate of the FWS-SF wetlands was initially controlled at about 0.62 m/day, and gradually increased to 2.82 m/day in the final culture phase. Solids in the recirculating water were efficiently removed through the FWS-SF wetlands by 69% (measured as total suspended solids) and 98% (measured as turbidity). Microbial mineralization and nitrification of the wetlands significantly reduced 5-day biochemical oxygen demand, total ammonia nitrogen and nitrite levels from the recirculating water by 53, 58, and 70 %, respectively. Nitrate was not prominently accumulated in the culture tank water during this study because denitrification occurred in the wetlands and exhibited an effective nitrate reduction (64%). At the end of the study, total weight of the harvested shrimp was 354 kg, and shrimp survival rate was estimated to be 72%. Accordingly, shrimp production per unit culture area was 5.5 kg/m^2, which is around 8 times higher than the production of conventional pond aquaculture. Constructed wetland was demonstrated to be a sustainable technology for regulating water qualities and saving water usage in a recirculating system for in-door intensive shrimp production.

Keywords Constructed wetland; recirculating aquaculture system; Pacific white shrimp; aquaculture wastewater

Introduction

Aquaculture is an important industry in Taiwan to supply seafood both for domestic demand and exporting trade. However, pond aquaculture requires a large amount of water resource and land area, and produces a polluted discharge, thereby creating various environmental impacts. Since most surface water bodies had been widely polluted, groundwater became the major water resource for aquaculture industry. Over withdrawal of groundwater has led to a considerable ground subsidence and seawater intrusion in several coastal areas of western Taiwan, where aquaculture intensively prevails. To reduce the impacts of aquaculture on environment, government promoted a Guiding Program for Aquaculture Industry in 1991 and a Ground Subsidence Control Program in 1995 (Hu, 1999). In these programs, development of recirculating aquaculture systems in fish and shrimp farms is one major and important approach to achieve sustainability in water usage for aquaculture.

A recirculating aquaculture system, integrating wastewater treatment process into aquaculture production to manage water quality, allows for intensive culture with limited pollutant discharge, thus reducing water and land usage, and minimizing adverse environmental impacts. Treatment units, such as sand filter, mechanical screen, submerged biofilters, rotating biological contactors, and/or fluidized bed reactors etc., are typically used in a recirculating aquaculture system (Tseng et al., 1998; Yang et al., 2001). These conventional methods can efficiently remove the pollutants and toxic materials from aquaculture wastewater. However, they require much higher capital cost and energy, depend on frequent maintenance, and produce sludge. These disadvantages have limited the will of farmers to use the recirculating aquaculture system instead of pond aquaculture. Accordingly, an effective and low-cost wastewater treatment process is imperative for sustainable development of aquaculture in Taiwan.

Previous studies have demonstrated that constructed wetlands are technically and economically feasible to remove the major pollutants from catfish, shrimp, milkfish, and rainbow trout pond effluents (Schwartz and Boyd, 1995; Sansanayuth et al., 1996; Lin et al., 2002a, 2002b; Schulz et al., 2003), and can act as a cost effective filter using in a recirculating aquaculture system (Zachritz and Jacquez, 1993; Tilley et al., 2002; Lin et al., 2003). However, the success of this approach in aquaculture production for field application is still seldom reported.

In this study, a commercial-scale recirculating aquaculture system was installed to demonstrate the effectiveness of constructed wetlands for managing water quality and controlling water usage, and the success in intensive shrimp production. The cost of the wetland based treatment process was also analysed to compare with a conventional treatment process.

Materials and methods

Recirculating aquaculture system

The recirculating aquaculture system was built in a fish farm at Tainan County, Taiwan during March 2001. This system mainly consisted of an indoor culture tank and an outdoor wetland based water treatment unit (Fig. 1), which was made of brick and concrete in main body. Pipelines made of polyvinyl chloride were installed to connect the culture tank and treatment unit for conveying the recirculating water.

The culture tank was 8 m × 8 m × 1.5 m (inner length, width and height) in size, maintaining a water depth of around 1.2 m. Tube diffusers, connected to an air compressor, were installed in the culture tank to supply oxygen for shrimp culture. Tank water flowed continuously to the water treatment unit by gravity, and the flow rate was controlled by a gate valve.

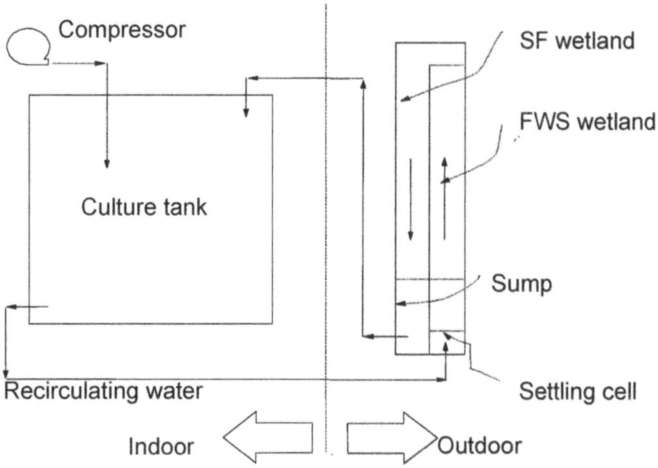

Figure 1 Schematic diagram of a recirculating aquaculture system integrating constructed wetlands

The water treatment unit was measured by 2.1 m × 15.2 m (inner width and length), and was divided longitudinally by a brick-concrete wall (10 cm thick) to form a U shape ditch with an equal inner width of 1 m. This ditch was separated into four cells: a settling cell, a free water surface wetland (FWS cell), a subsurface flow wetland (SF cell), and a sump.

The settling cell (inner length of 1.5 m) received the recirculating water from the culture tank. A perforated acrylic baffle was installed across the wide direction of the cell at 25 cm from the inlet end for flow distribution. The bottom of the cell sloped at about 14% towards the inlet end, with a sludge hopper located at the bottom of the inlet end. Sludge was occasionally siphoned out from the hopper. The deepest water depth was around 1.2 m.

The effluent from settling cell flowed through a perforated full-width brick-concrete wall into the FWS cell. The FWS cell (inner length of 12 m) contained a 30 cm layer of local soil (Jender silt loam, 25°37'N, 166°55'E) at the bottom and 40 cm of averaged water depth above the soil layer.

The effluent of the FWS cell entered the SF cell through a perforated full-width brick-concrete wall. The SF cell (inner length of 15.7 m) included 80 cm of river gravel (nominal diameter 10 to 20 mm), providing a porosity of 45%, and 65 cm of averaged subsurface water flow within the gravel layer. A lateral perforated manifold, as a collection drain, was installed at the bottom of outlet end of the SF cell. The manifold was extended to the sump (inner length of 1 m) and connected with a 90^0 elbow to control the water depth of wetland cells. After re-aeration in the sump, the wetland treated effluent was recycled to the culture tank by a submerged pump.

Cattail (*Typha angustifolia* L.) and common reed (*Phragmites australis*) were planted in the FWS and SF cell, respectively, both with initial density of around 6 plants/m^2 in the late March 2001.

Shrimp culture conditions

Shrimp culture was conducted during a warm season from 13 June to 17 October 2002. Postlarvae (PL) of Pacific white shrimp (*Litopenaeus vannamei*) with initial body weights of 5 to 6 mg/PL were introduced into the culture tank with stocking densities of around 625 PL/m^2. A brackish groundwater containing approximately 0.3% salinity was used for shrimp culture in this study. The shrimp were fed with a powder diet during the initial four weeks of culture and a pellet diet after the four weeks of culture. These commercial diets contained around 45% protein, 2 % fat, 3% cellulose, 17% ash, and 11% water. The feeding rate was adjusted according to the intake rate of the shrimp. The shrimp were fed manually at around 8 A.M. and 4 P.M. each day. After four weeks of culture, 30 shrimps were taken from the culture tank every two weeks, and their body weight and length were measured to estimate the shrimp growth.

Water sampling and analysis

Water samples were taken twice a week from outflow end of the culture tank, influent of the FWS cell, effluent of the FWS cell, and effluent of the SF cell. Such sampling was usually performed at around 10 A.M. on each sampling date. The samples were analyzed for total suspended solids (TSS) and turbidity. The filtrates left by filtering water samples through glass-fiber filters were measured for 5-day biochemical oxygen demand (BOD_5), total ammonium nitrogen (TAN), nitrite nitrogen (NO_2-N), nitrate nitrogen (NO_3-N), and orthophosphate (PO_4-P). Temperature, pH and dissolved oxygen (DO) of water in the sampling locations were also monitored. All the analytical measurements were carried out according to Standard Methods (American Public Health Association, 1995).

Tests for significant difference in water quality between influent and effluent of the treatment cells were determined by paired *t*-tests at a significance level of 0.05 (OriginLab, 1996) for each set of data.

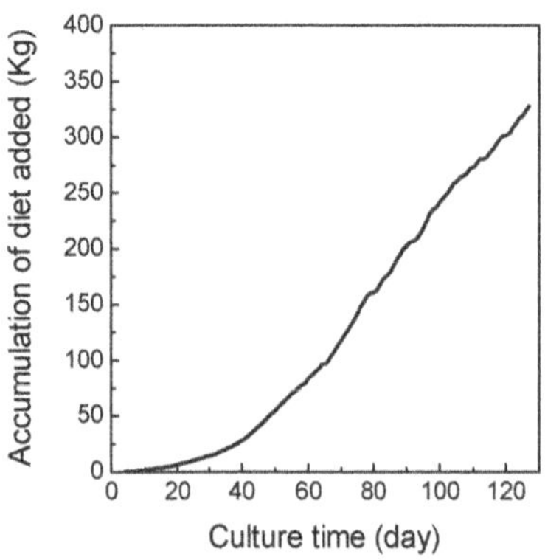

Figure 2 The increasing curve of shrimp diet added in culture tank of the recirculating aquaculture system

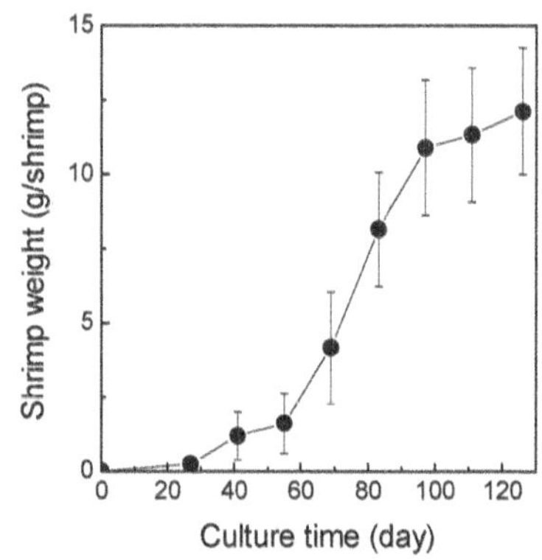

Figure 3 Growth curve in term of weight of shrimp in the recirculating aquaculture system

Results and discussion

Shrimp culture and growth

The increasing curve of shrimp diet added was found corresponding with shrimp growth curve (Figures 2 and 3). Shrimp grew slowly in the first 30 days, exhibiting an initial growth phase. Shrimp growth rate gradually increased and reached a rapid growth phase during days 55 to 95 (Figure 3). A declining growth phase was observed afterward because crowding effect due to intensive culture. Throughout the whole study of 126 days culture, the shrimp diet totalled to 328 kg and final shrimp body weight reached to a food size of 12.3 g in average. Maximum growth rate of shrimp was recorded to be 2 g/shrimp/week, with an average growth rate of 0.1 g/shrimp/day.

Recirculating water control

Flow rate of recirculating water into the treatment unit was initially controlled at about 17.3 m^3/day (equivalent to a hydraulic loading rate of 0.62 m/d for the FWS-SF wetlands), and gradually increased to 78 m^3/day (hydraulic loading rate of 2.82 m/d) in the final culture stage (Figure 4). This flow control strategy was made to correspond with the tendencies of diet added (i.e., pollutant loaded) and shrimp growth with culture time. Small flow rate would be sufficient to handle the pollutants generated in the initial culture phase, with also saving the pumping cost for recirculating water.

Performance of constructed wetlands and water quality

Prior to this study, two phases of culture using this same recirculating system had been carried out to collect data for determining the wetland area required for management of water quality in the intensive shrimp aquaculture from April to June 2001 and from August 2001 to January 2002, respectively (Lin et al., 2005). This study is indeed the third phase of culture. Macrophytes in constructed wetlands had reached a stable density more than 90 plants/m^2 since the end of the second phase. Macrophytes were not purposely harvested except dead plant detritus were occasionally removed to avoid a blockage problem of water flow. Continuous-flow operation of the constructed wetlands was ceased during the period between two phases, but still maintaining the water depth to sustain the growth and survival of both macrophytes and microorganisms in wetland cells.

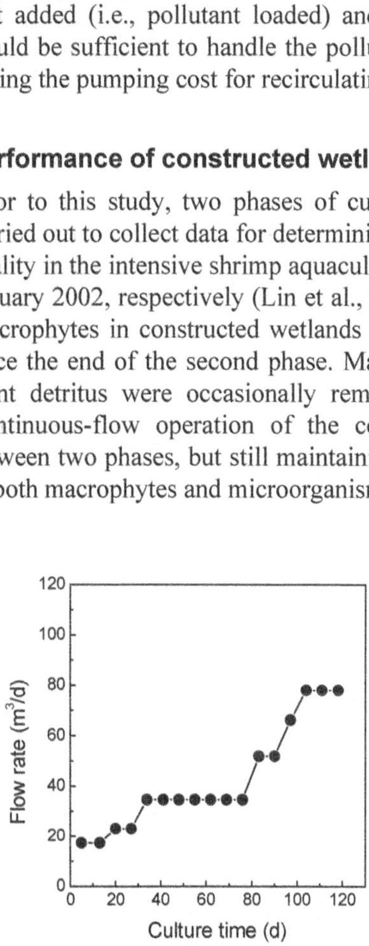

Figure 4 Flow rate of recirculating aquaculture water through constructed wetland system during shrimp culture

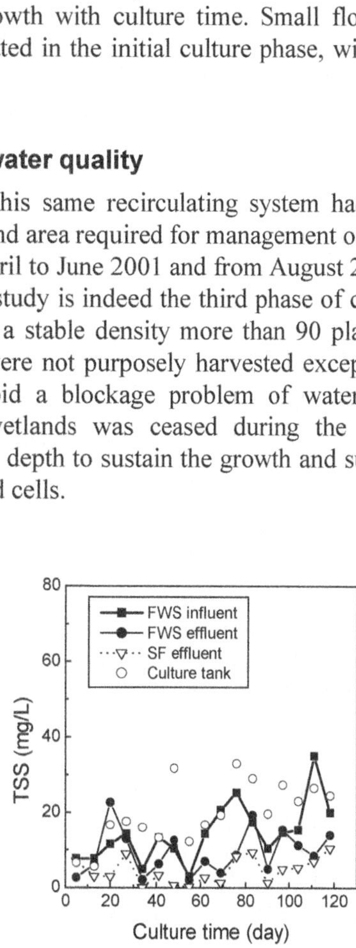

Figure 5 Time course of TSS at various sampling positions in the recirculating aquaculture system

Table 1 Treatment results (mean ± standard deviation) for various parameters of water quality by the FWS-SF wetlands (n = 18)

Parameter	FWS influent	FWS effluent	SF effluent	Average removal efficiency (%)	Culture tank
pH	8.2±0.6	8.1±0.5	7.8±0.5	-	8.1±0.2
Alkalinity (as $CaCO_3$ mg /L)	347±23	342±18	334±16	-	345±22
DO (mg/L)	4.7±2.4	2.4±2.7	0.56±0.65	-	6.6±0.4
TSS (mg/L)	14.5±7.8	9.5±6.1	4.6±3.4	69±23	20±8
Turbidity (NTU)	1.1±0.9	0.35±0.71	0.01±0.02	98±2	1.9±1.2
BOD_5 (mg/L)	6.6±4.9	5.3±4.1	2.7±1.9	53±25	7.3±5.9
TAN (mg/L)	0.58±0.65	0.64±0.62	0.20±0.25	58±25	0.37±0.24
NO_2-N (mg/L)	0.48±0.47	0.37±0.37	0.08±0.09	70±32	0.43±0.44
NO_3-N (mg/L)	2.79±2.9	2.92±2.95	1.98±2.73	64±35	3.13±3.14

There was a slight decrease of pH and alkalinity in recirculating water flowing through the FWS-SF wetlands (Table 1). This decrease might be due to nitrification occurring in the treatment wetlands. However, pH of water in the culture tank was stably maintained at a suitable range of 7.7~8.5 without any pH adjustment. Solids in the recirculating water were efficiently removed through the FWS-SF wetlands by 69% (measured as TSS reduction) and 99% (measured as turbidity reduction), leading to a tank water of 20±8 mg/L for SS and 1.9±1.2 for NTU (Figure 5 and Table 1). Solids in the recirculating aquaculture water are mostly generated from feed residue and shrimp excreta. They can be removed by constructed wetlands via sedimentation and filtration that occur within plant stems and roots and within the gravel that forms the substratum. Organic solids can be further stabilized in wetlands through biological processes such as microbial degradation (Lin et al., 2002b). Thus, solids accumulation in wetland cells was ignored during this study, and processing of sludge produced from wetlands was not required.

Microbial mineralization and nitrification occurring in the treatment wetlands significantly (p<0.05) reduced BOD_5, NH_4-N and NO_2-N levels in the influent water by 53, 58, and 70 %, respectively, and resulted in consistently low levels of BOD_5 (7±6 mg/L), TAN (0.37±0.24 mg/L), and NO_2-N (0.43±0.44 mg/L) in culture tank water (Table 1, Figures 6–8). These levels were well below the safe levels for rearing *L. vannemei* juveniles (2.44–3.95 mg/L for TAN and 6.1–25.7 mg/L for NO_2-N) reported by Lin and Chen (2001, 2003). Moreover, the simultaneous reduction of TSS and BOD_5 from the influent to effluent indicates that degradation and stabilization of solids actually occurred in the constructed wetlands. The results of transitions of influent-effluent water quality (Figures 5–9) suggests that the constructed wetlands required a brief start-up period, less than 10 days, to achieve a stable and consistent removal of major pollutants when initiating a new culture phase.

Although nitrification efficiently occurred in wetlands, NO_3-N was not accumulated in the SF effluent. On the contrary, NO_3-N concentrations of the SF effluent were always lower than those of the FWS influent in 83% of samples (13 of 16) (Figure 9). This result led to 64% of mean nitrate reduction from FWS influent to SF effluent (Table 1). These observations might be caused by the denitrification performance of the constructed wetlands. Consequently, NO_3-N in culture tank water exhibited a low level of 3.13±3.14 mg /L during this study (Table 1). The ability of wetland denitrification here is shown more effective than that in the previous two phases of culture (Lin et al., 2005), probably due to a denser macrophyte achieved in this study providing more available organic carbon to boost denitrification.

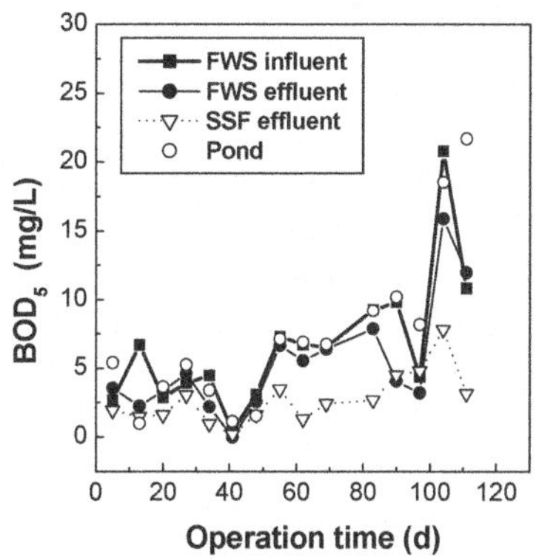

Figure 6 Time course of BOD_5 at various sampling positions in the recirculating aquaculture system

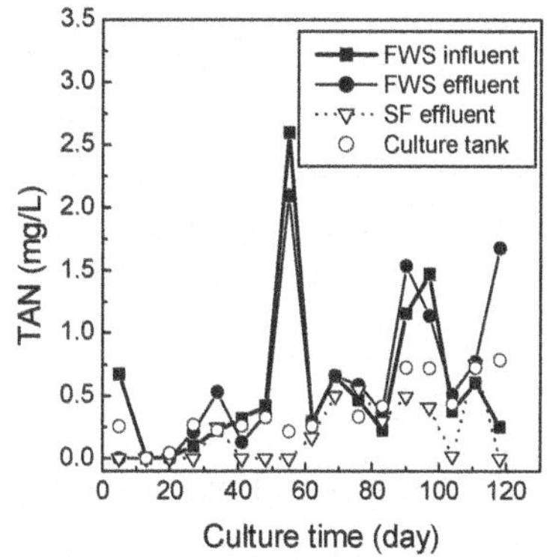

Figure 7 Time course of TAN at various sampling positions in the recirculating aquaculture system

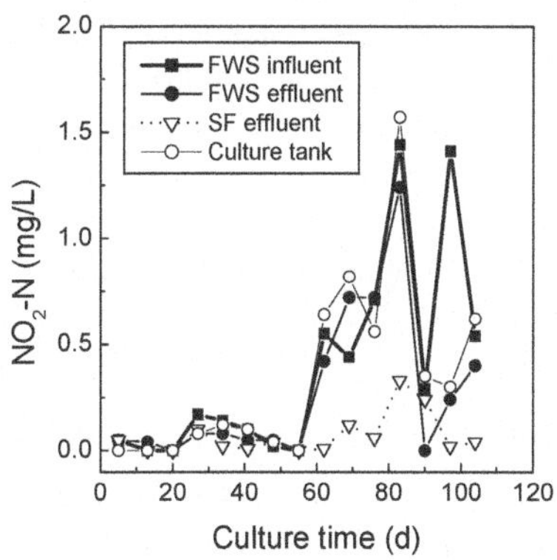

Figure 8 Time course of NO_2-N at various sampling positions in the recirculating aquaculture system

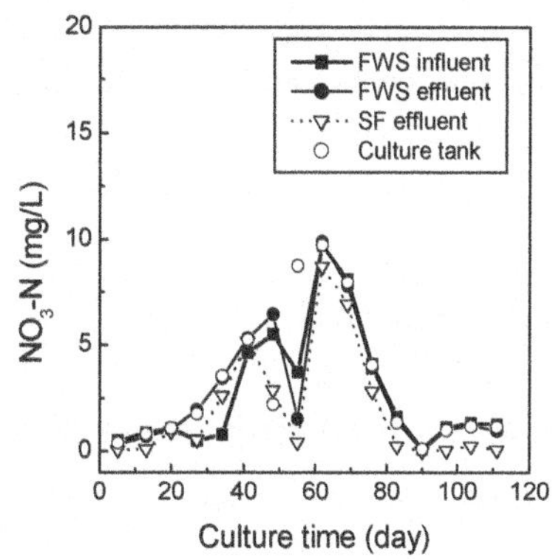

Figure 9 Time course of NO_3-N at various sampling positions in the recirculating aquaculture system

Table 2 Capital cost analyses of the wetland based treatment unit for an indoor recirculating aquaculture system

Description	Cost (US$)	Unit Cost (US$/m^2)	Cost (%)
Basins construction	7,900	39.5	65.6
Gravels and soil	2,670	13.35	22.1
Planting	270	1.35	2.2
Electric, pumping and pipe systems	1,200	6	10.1
Total	12,040	60.2	100

Basis: 200 m^2 scale concrete made basin.

The performance of treatment wetlands in this study is easily comparable to that of the conventional treatment process (including micro-screen, foam fractionator, and biological filter) employed in a recirculating shrimp production system reported by Davis and Arnold (1998), in which TAN and NO_2-N were maintained at the range of 0.36~1.16 and 0.30~1.48 mg/L, respectively, in culture chamber.

Shrimp production and water usage

At the end of the study, total weight of the harvested shrimps was 354 kg, and shrimp survival rate was estimated to be 72%. Overall food conversion ratio, defined as the ration of the total amount of diet added to the increment of total weight of shrimps harvested, was estimated to be 0.92. This efficient conversion of food to shrimp growth might be resulted from a suitable water quality and health environment managed by constructed wetlands. Accordingly, shrimp production per unit culture area was 5.5 kg/m^2, which is around 8 times higher than shrimp production in conventional earthen pond aquaculture. This shrimp production is rather successful in comparison with an indoor recirculating white shrimp culture system employing conventional treatment processes (Lin et al., 2000), in which a higher stock density (1300 PL/m^2) with a higher initial weight (66 mg/PL) was used, and achieving a food size of 10~11 g/shrimp and a production of 6.7 kg/m^2 after 105 days culture.

No water discharge or displacement occurred during this study, except for replacing water lost through evaporation.

Cost analysis for a wetland based treatment unit

Capital cost analyses of the wetland based treatment unit for an indoor recirculating aquaculture system is provided in Table 2. These costs were estimated base on a 200-m^2-scale concrete made treatment unit scaling up from the results of this study. Basin construction and gravel are the two major cost items, accounting for 65.6 and 22.1 % of the total capital cost, respectively. The capital cost will be further reduced if treatment wetlands are built of earthen basin and gravel is substituted by recycling substrates.

Chen et al. (2002) analysed cost and benefit of a 0.1 ha automatic indoor recirculating shrimp culture system with 420 m^2 required for culture tank. In that system, capital cost of the conventional wastewater treatment process used (including micro-screen, biological filter, electrical disinfection device) was approximately US$71,700 and contributed to the most significant part (34%) of the total capital cost. If constructed wetland technology is applied to such scale of shrimp culture (*i.e.*, 420 m^2 for culture area), a land area of 200 m^2 will be required for construction of the wetland based treatment unit (treatment wetland plus a settling basin), according to an area ratio (0.47) of treatment unit to culture tank used in this study. As a consequence, the capital cost of the wetland based treatment unit for an indoor recirculating shrimp culture system will be US$12,040 that is less than 20% of the conventional wastewater treatment process. Operation and maintenance costs of the wetland based treatment unit will mainly include electricity demand for pumping the recirculating water.

Conclusions

Recirculating aquaculture systems incorporating conventional mechanical treatment processes have been development for decades for sustainability in aquaculture. However, the systems are not widely used probably because they have not been cost effective for large-volume culture systems. This study demonstrated that a wetland based treatment technology, mainly including two types (free water surface and subsurface flow) of constructed wetland, is an economic and efficient approach to manage the water usage and water quality for intensive aquaculture production. Wetland treatment efficiently and simultaneously removed TSS (69%), turbidity (98%), BOD_5 (53%), TAN (58%), NO_2-N (70%), and even NO_3-N (70%) from recirculating water, leading to an appropriate and non-harmful water quality to aquaculture. The culture system of this study gained a 5.5 kg/m^2 of shrimp production, which is comparable to the culture system using a conventional treatment process and is 8 times higher than the production of earthen pond aquaculture. On the basis of serving equal culture scale, the capital cost of a wetland based treatment unit is only about 20% of a conventional treatment process. Accordingly, incorporating constructed wetlands technology into a recirculating aquaculture system will significantly reduce the overall cost and enhance sustainability of the system.

References

American Public Health Association (1995) Standard methods for the examination of water and wastewater. APHA, AWWA, and WPCF, Washington, DC.

Chen, S., Liang, R.Y., Lia, K.H. and Lin, N.H. (2002) Research and development of an automatic indoor recirculating shrimp culture system. *China Fisheries Monthly* **598**, 37-49.

Davis, D.A. and Arnold, C.R. (1998) The design, management and production of a recirculating raceway system for the production of marine shrimp. *Aquacult. Eng.* **17**(3), 193-211.

Hu, X.H. (1999) Aquaculture in Taiwan. *Fish. Ext.* **157**, 13-26.

Lin, Y.C. and Chen, J.C. (2001) Acute toxicity of ammonia on *Litopenaeus vannamei* (Boone) juveniles at different salinity levels. *J. Exp. Mar. Biol. Ecol.* **259**, 109-119.

Lin, Y.C. and Chen, J.C. (2003) Acute toxicity of nitrite on *Litopenaeus vannamei* (Boone) juveniles at different salinity levels. *Aquaculture* **224**(1-4), 193-201.

Lin, M.N., Tseng, B.S, Chiu, C.S., Ting, Y.Y. Chen, S., Liang, R.Y. and Lia, K.H. (2000) Culture of white-leg shrimp, *Penaeus vannamei* in multi-layered and automatic indoor system in autumn, with a stocking density of 1,300 individuals or 2,167 individuals per cubic meter. *Bulletin of Tainan Fish Culture Station* **4**, 1-13.

Lin, Y.F., Jing, S.R., Lee, D.Y. and Wang, T.W. (2002a) Nutrient removal from aquaculture wastewater using a constructed wetlands system. *Aquaculture* **209**(1-4), 169-184.

Lin, Y.F., Jing, S.R., Lee, D.Y. and Wang, T.W. (2002b) Removal of solids and oxygen demand from aquaculture wastewater with a constructed wetland system in the start-up phase. *Water Environ. Res.* **74**(2), 136-141.

Lin, Y.F., Jing, S.R., Lee, D.Y. and Wang, T.W. (2002c) Effect of macrophytes and external carbon sources on nitrate removal from groundwater in constructed wetlands. *Environ. Pollut.* **119**(3), 413-420.

Lin, Y.F., Jing, S.R. and Lee, D.Y. (2003) The potential use of constructed wetlands in a recirculating aquaculture system for shrimp culture. *Environ. Pollut.* **123**(1), 107-113.

Lin, Y.F., Jing, S.R., Lee, D.Y., Chang, Y.F., Chen, Y.M. and Shih, K.C. (2005) Performance of a constructed wetland treating intensive shrimp aquaculture wastewater under high hydraulic loading rate. *Environ. Pollut.*, **134**, 411-421.

Sansanayuth, P., Phadungchep, A., Ngammontha, S., Ngdngam, S., Sugasem, P., Hoshino, H. and Ttabucannon, M. S. (1996) Shrimp pond effluent: pollution problems and treatment by constructed wetlands. *Water Sci. Technol.* **34**(11), 93-98.

Schulz, C., Gelbrecht, J. and Rennert, B. (2003) Treatment of rainbow trout farm effluents in constructed wetland with emergent plants and subsurface horizontal water flow. *Aquaculture* **217**(1-4), 207-221.

Schwartz, M.F. and Boyd, C.E. (1995) Constructed wetlands for treatment of channel catfish pond effluents. *Progr. Fish-Cult.* **57**(4), 255-267.

Tilley, D. R., Badrinarayanan, H., Rosati, R. and Son, J. (2002) Constructed wetlands as recirculation filters in large-scale shrimp aquaculture. *Aquacult. Eng.* **26**(2), 81-109.

Tseng, K.F., Su, H.M., Su, M.S. (1998) Culture of Penaeus monodon in a recirculating system. *Aquacult. Eng.* **17**(2), 138-147.

Yang, L., Chou, L.S. and Shieh, W.K. (2001) Biofilter treatment of aquaculture water for reuse applications. *Wat. Res.* **35**(13), 3097–3108.

Decolourising yellow water for its reuse in sustainable sanitation

F. Tettenborn, B. Lindner, J. Behrendt and R. Otterpohl

Institute of Wastewater Management, Hamburg University of Technology (TUHH), D-21071 Hamburg, Germany. (E-mail: tettenborn@tuhh,de; b.lindner@tuhh.de; behrendt@tuhh.de; otterpohl@tuhh.de)

Abstract For a successful realisation of many concepts, in which parts of yellow water are used as process water, an effective decolourisation water is essential. Thus, the effect of ozone and UVC-radiation on the colour of yellow water was investigated. Samples with different concentrations of yellow water and after different pre-treatment steps were treated in an ozonisation lab-scale plant and in an UVC-radiation reactor and examined before, and during treatment and storage. Both processes were successful in producing a lasting removal of colour. Energy balance showed that ozone treatment was much more efficient. The great advantage of UVC-radiation is the very simple and safe application. In addition to colour removal, ozone contributed to deodorization as a consequence of ammonia and phenol oxidation. Special attention was given to colour evaluation and computation. Knowing the cause of colour appearance in urine, models were developed describing possible ways of destruction of the colour causing molecules.

Keywords Black water cycle; decolourisation; ozone; pigments in urine; UVC-radiation; yellow water

Introduction

Urine treatment and usage is becoming more and more important in sustainable sanitation concepts. Both, dissolved nutrients as well as the water fraction offer a variety of new solutions in reuse and closed loop concepts. Fertiliser for agricultural use, industrial resources (e.g. ammonium sulphate, ammonia water) and process water are the substances that can be of economic interest. However, in some systems, in which yellow water is used as process water (e.g. black water cycle), characteristics as odour or colour are not wanted. Thus, processes were investigated with the goal of a lasting colour removal.

In the past several oxidation and adsorption processes (AOP, H_2O_2, UV, activated charcoal) (Gulyas *et al.* 2004) did not show satisfying results in colour reduction in urine. Hence, colour removal was investigated using high frequent UVC-light (185 nm) and ozone.

Possible use of urine

The use of urine as fertiliser in agriculture is widely documented (Jönsson 2002; Vinneras *et al.* 2004). Urine was also used in the past for many industrial processes (Lange and Otterpohl 2000). E.g. in combustion ammonia is used for NO_x removal in flue gas.

New innovative concepts are necessary for saving future drinking water supplies and for recovering nutrients. Closed loop systems, such as the black water cycle (BWC, explained below), offer a promising perspective. Based on source separation little to no fresh water is needed, while recovering nutrients for fertiliser production and organics for soil conditioning is possible. For acceptance and a broad adoption of such a closed loop system, the applied water has to be absolutely clear and colourless. An effective decolourisation of the circulating flushing liquid in the BWC is essential for a successful implementation.

Black water cycle

With an extremely small water consumption nutrients and organic matter from domestic wastewater can be recovered and made accessible to the natural food cycle. This way fresh water occurrences are pre-served, so that potable water sources are guaranteed in the long run; nutrients and organic matter from the food cycle are made re-accessible for it, so that future expensive, and resource-wasting production of artificial fertilisers can be evaded; finally, environmental pollution by sewage discharge is avoided. Another big advantage of this system is the fact that also in arid areas conventional water flushing toilets can be used without effecting the lasting water resources. Thus an enormous improvement of sanitary conditions is reached, which has direct impact on the quality of life and the health situation of water-poor countries.

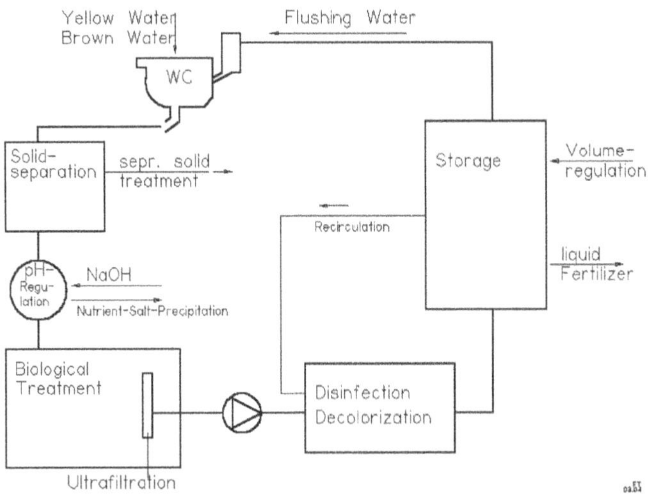

Figure 1 Concept of Flushing water cycle

The concept, patented by U. Braun and further investigated at TUHH by B. Lindner, is based on a separation of black and grey water. While the organic matter of black water can be converted to soil conditioner (composting/vermicomposting) and/or energy (anaerobic digestion), the liquid fraction is conditioned in a way, that it can be reused as flushing liquid. Over time part of the accumulated liquid can be extracted and be used as liquid fertiliser. The circulating liquid fraction has to be hygienically safe, colour- and odourless, and consistent for storage without building residues.

After the first treatment steps including ultra filtration the remaining liquid consists mainly of biologically pre-treated diluted yellow water, containing dissolved salts, some organic molecules, and very few viruses. The liquid is further disinfected and colourless, so that it can be reused as flushing water.

Urine, some characteristics and components

The composition of urine is very complex and varies from person to person. The colour is caused by bile pigments, breakdown products of haemoglobin, particularly urobilin which is formed from colourless urobilinogen. Urobilinogen, that is present in fresh urine will get transformed to yellow urobilin over time, which explains the increase in colour of urine during storage. Normally there is no bilirubin contained in urine. The amount of urobilinogen in urine is around 2–17 mg/l (CIBA Geigy AG). From the observation of colour change during the experiments it could be derived that in urine the amount of contained urobilin is about 1/3 of the amount of urobilinogen.

Other substances that should be mentioned because of their quantities are urea, different amino acids, and a variety of salts. COD is about 7 g O_2/l.

Table 1 Some components contained in urine, units in g/(d*pers) (CIBA Geigy AG).

N_{tot}	~10	Potassium (K^+)	~2.7	Calcium (Ca^{++})	~0.2
Chloride (Cl^-)	~5	Phosphate (PO_4^{3-})	~1–5	S_{tot}	~1.3
Sodium (Na^+)	~3–5	Magnesium (Mg^{++})	~0.15	pH	5–8

Colour discussion in general

Detection of colour by eye-sight is a very subjective topic. Thus the transmitted spectral light was measured by spectrophotography. From this colour was expressed according to the CIELab-system. In cylindrical coordinates chroma C^*_{ab} describes the intensity, the hue angle h_{ab} specifies the colour itself, while L* quantifies lightness. An increase of the colour intensity will lead to a decrease of lightness. Equivalent, the calculated value of the colour can increase, when lightness is decreasing.

If the spectral curve cannot be surely predicted, or if a precise value for colour is necessary, the whole spectral bandwidth of visible light has to be recorded.

Yellow water absorbs the violet/blue part of the spectral light. This is because of the energy absorption of the conjugated π-electron system from the two middle pyrrol rings of urobilin. The angle hue is changing only to a very small extend. L* is affected strongly by turbidity. An increase of turbidity leads to a decreasing lightness. In this study turbidity is expressed as the difference between absolute lightness (L*=100) and the relative lightness L*. A correlation with standard turbidity measurements was not conducted.

The factor that is describing the intensity of colour is chroma C*. Concentrated fresh urine has a chroma of about C*=20. Biologically pre-treated yellow water was at a 30% dilution similar to fresh urine at a dilution of 40%, C*=8. At a dilution of 10 % slight yellow colour was still detectable by eye sight (C*=3). A dilution of 5 % led to C*=1, where no yellow colour was detectable by eyesight.

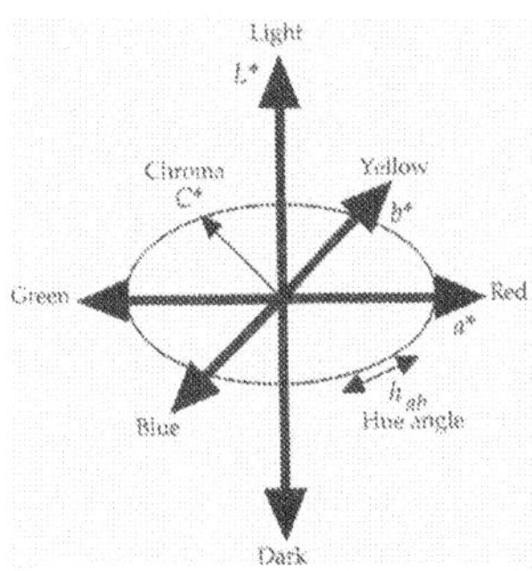

Figure 2 Scheme of CIELab-system

Ozonisation and UVC-radiation

Ozone is highly reactive. Materials/molecules that have chemical double bonds, or that contain oxidisable elements can be oxidised by ozone. The specific and direct reaction of ozone with organic substances is due to an electophilic addition that leads to a splitting of the C=C double bonds (Crigée mechanism). Products can be among others aldehyde, ketone, carbon acids (Bünning), and hydrogen peroxid (Gulyas 2000). Reaction rate constant is rather slow in the range of k_D=1.0–10^3 $M^{-1}s^{-1}$ (Gottschalk). During indirect reaction secondary oxidants, especially hydroxyl radicals (OH°), are formed by the decay of ozone. These react non-selectively and immediately (k=10^8–10^{10} M^{-1} s^{-1}) with solutes. Energy consumption during the experiments was about 6 Wh/g O_3. For the production of oxygen a value of 10 Wh per g O_3 was used in the energy balance.

UVC-radiation (wavelength between 100 and 280 nm) is very energy intense. At 185 nm free hydroxyl radicals are formed directly from water molecules, hydrogen peroxide (H_2O_2) is a by-product of this process. At 254 nm H_2O_2 will decay into radicals. Below 240 nm ozone is produced out of dissolved oxygen. Ozone will also decay into OH° radicals. Besides of these chemical reactants different photochemical processes are caused by UVC-radiation.

Analyses

The spectral curve, needed for colour evaluation, was measured using a Jasco spectrophotometer V-550. For pH and conductivity WTW probes were used. TOC, IC, and TN was analysed according to German standard methods. Ammonium, nitrite, and ozone were monitored by quick tests, respectively ozone analysers.

Research conduction

Both lab-scale plants were designed as continuous flow reactors with a circulating substrate. The flow rate in the ozone reactor (bubble column volume: 300 ml, diameter: 2–4.5 cm) was 1.2 l/h, ozonisation rate between 0.33 g O_3/h and 2.4 g O_3/h.

The reaction surface of the UVC-reactor was 224 cm^2 with an energy input of 80.4 mW/cm^2. The flow rate was 12 l/h, duration of radiation 13.5 min/h.

Experiments were conducted using fresh urine and later biologically pre-treated urine. For ozonisation and UVC-radiation always the same starting solution was used.

Samples were taken during the active phase of ozonisation and UVC radiation. Subsequently all samples were examined during a storage phase of up to 4 weeks.

Results

Active phase, ozonisation, and UVC-radiation of fresh untreated yellow water (40%)

Ozonisation of fresh untreated yellow water (40% dilution) led to a fast decolourisation. After 50 minutes of ozonisation the intensity of yellow colour was reduced by 50%. After 3 hours 20 minutes absolutely no colour was detectable any more. The pH dropped from pH 6.5 in the beginning to finally pH 3.6 after 24 hours of ozonisation. A small TOC increase, going along with a slight IC decrease, could be observed during ozonisation. TN remained stable. Odour was depleted already after 50 minutes. After long ozonisation (24 hours) a smell of vinegar could be observed.

During UVC-radiation of fresh yellow water an increase in the intensity of yellow colour could be observed over the first 26 hours. Afterwards the intensity of colour was reduced slowly down to C*=3.8 after 7.5 days, what was colourless by eye sight compared to earlier samples. Turbidity was quite stable at 100–L*=4. The pH increased in the beginning faster. Starting from pH 6.8 to finally pH 8. TOC:IC varied from 62:38 over 55:45 to finally 90:10. Odour changed from fresh urine to the dull odour of amine.

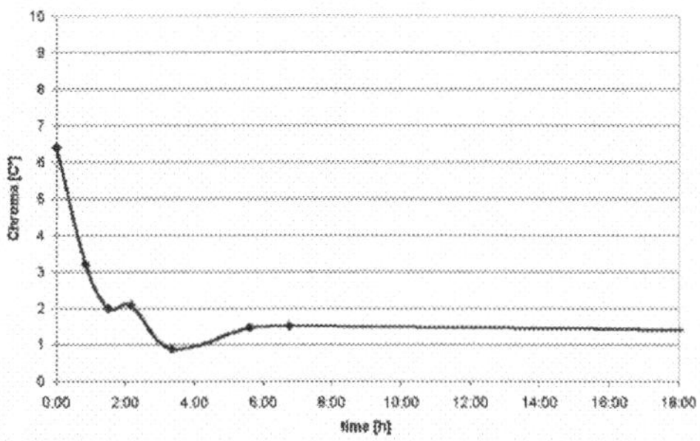

Figure 3 Change of chroma during ozonisation of fresh yellow water (40%)

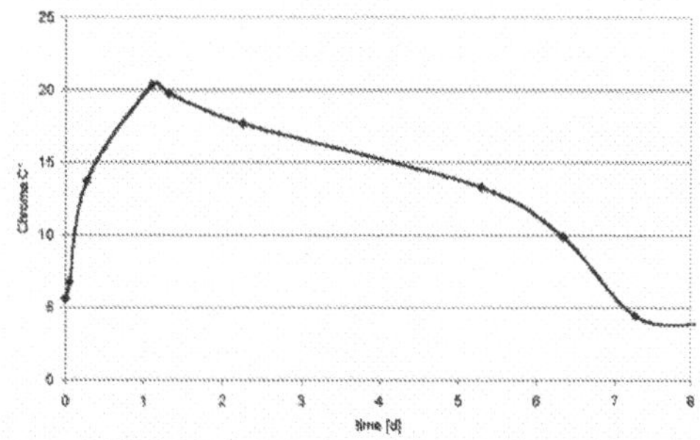

Figure 4 Change of chroma of fresh yellow water (40%) during UVC-radiation

Active phase, ozonisation and UVC radiation of pre-treated yellow water (40%)

The colour reduction in biologically pre-treated yellow water (40% dilution) could be reached already with very low ozone doses (0.3 g O_3/h). After 2 hours no colour could be detected any more (C*=0.9). Chroma was further reduced down to C*=0.4. There was hardly any turbidity (100–L*=0.3). pH remained stable at 8.3. There was a slight change in ratio TOC:IC from 70:30 to 40:60 in the end. TN remained stable. Already in the beginning, there was hardly any odour detected. Towards the end it became more “fresh”.

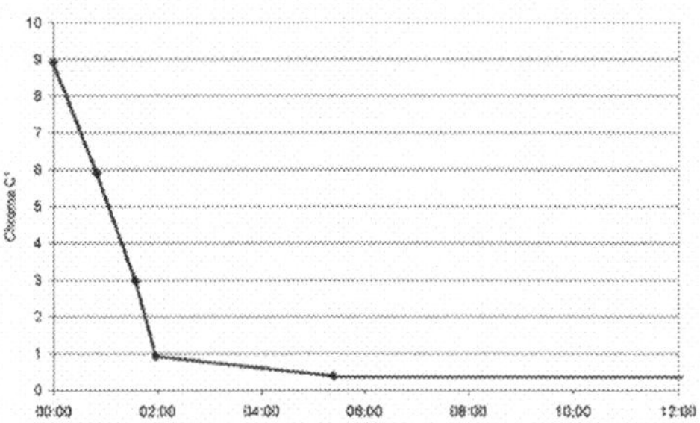

Figure 5 Change of chroma of pre-treated yellow water during ozonisation

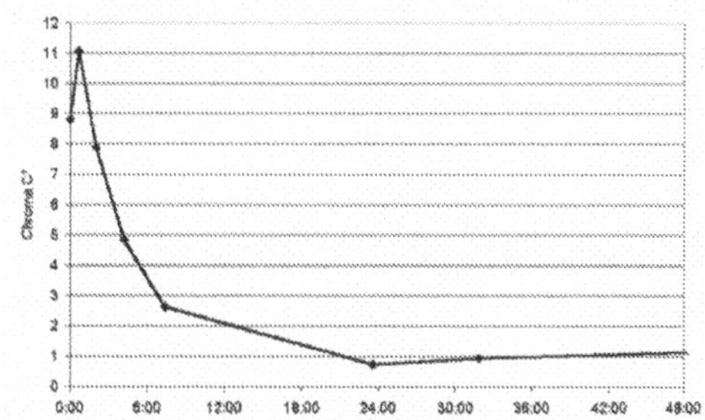

Figure 6 Change of chroma of fresh yellow water (40%) during UVC-radiation

During the first hour of UVC-radiation of biologically pre-treated yellow water there was a slight increase in chroma from C*=8.8 to C*=11. After 24 hours of UVC-radiation no colour could be detected (C*=0.74). There was no turbidity at any time. The value 100-L* even decreased during UVC treatment from 2.7 to 100–L*=0.2. There was a pH reduction from pH 8.3 in the beginning to finally pH 5.75 after 72 hours of UVC radiation. TOC, IC, and TN seemed to remain stable. After long UVC radiation of the pre-treated yellow water a slight odour reminded on solvents, or amine.

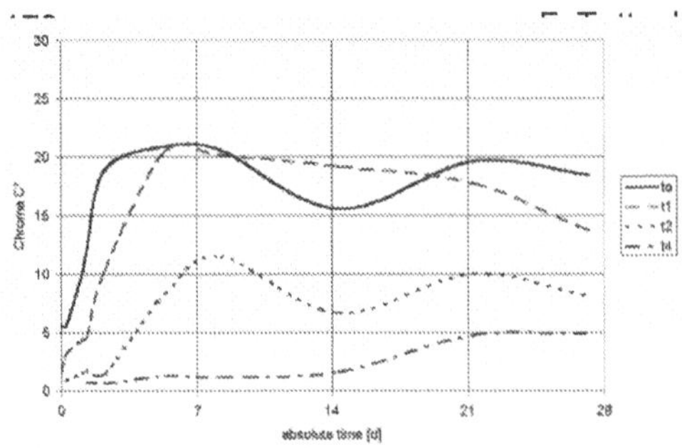

Figure 7 Chroma-change of samples during storage after ozonisation

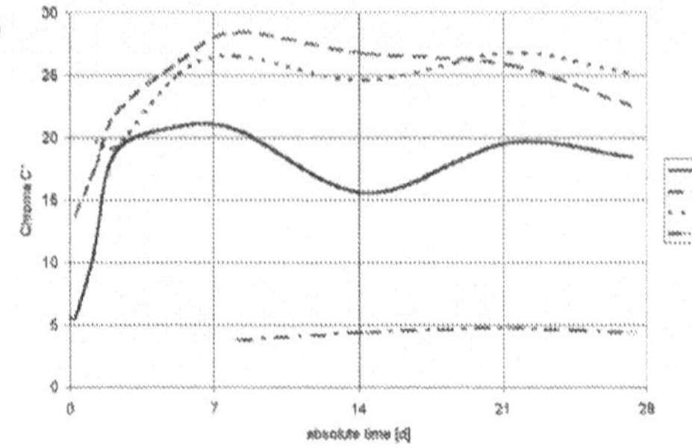

Figure 8 Chroma-change of samples during storage after UVC-radiation

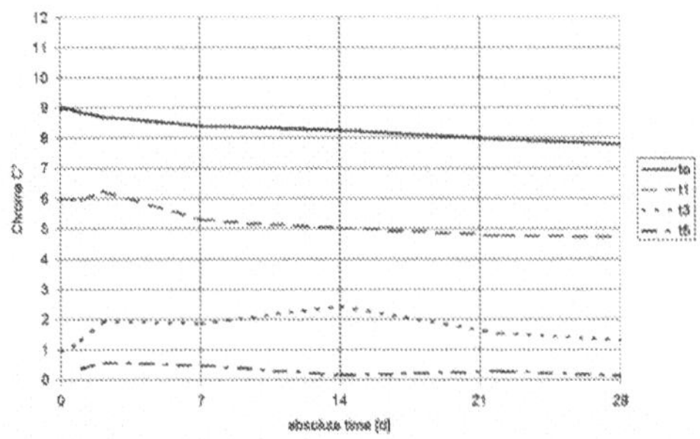

Figure 9 Chroma-change of samples during storage after ozonisation

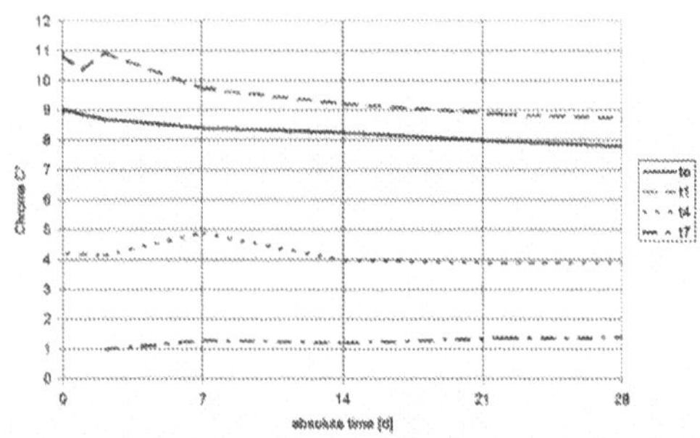

Figure 10 Chroma-change of samples during storage after UVC-radiation

Storage of untreated urine

During storage of untreated, diluted (40%) fresh urine (plot t0 in Figures 7 and 8) the intensity of yellow colour raised within 54 hours from C*=6.2 to a maximum of C*=20. During storage for the next 24 days the intensity of yellow colour changed from C*=16 to C*=19. Turbidity increased in the beginning up to 100–L*=28.1 at the climax of colour intensity. After 20 days it was reduced again to 100–L*=8.8. pH increased during storage slightly from pH=6.5 to pH=9.1 after 9 days where it remained. Two thirds of TOC was converted to IC after 27 days.

Storage phase after treatment of initially fresh yellow water

Samples were stored over a period of 27 days. After a total ozone dose of 2.2 gO_3/(l yellow water) (tl in Figure 7) the rise of chroma took place with a small delay. After 5.5 days a maximum of C*>20 was reached. After ozone dosage of 12 gO_3/(l yellow water) (t2 in Figure 7) the low chroma value remained for two days. In the following days chroma increased. Even after an ozone dose of 48 gO_3/(l yellow water) (t4 in Figure 7) chroma started to increase again after 15 days up to finally C*=5. pH increased in all samples, after higher ozone doses with a delay of 1–14 days. Difference between starting and end pH was around 3. Also odour of stored urine could be detected accordingly to an increase in chroma.

During UVC-radiation of formerly fresh urine chroma increase was intensified. Maximum chroma after radiation doses of 39 W h (t2 in Figure 9) up to 325 W h (t4 in Figure 9) was higher than in the initial sample after an equivalent time. Chroma values remained after reaching their maximum. Only at very high UVC radiation doses (sample t9 in Figure 8 was treated with 756 W h of UVC radiation) values remained stable over storage time, in this case below C*=5.

Storage phase after treatment of pre-treated yellow water

After treatment of biologically pre-treated yellow water with ozone, respectively with UVC radiation: all values remained very stable. There was hardly any change in chroma, pH and other monitored values.

Discussion and conclusion

Earlier investigations (Gulyas *et al.* 2004) showed that pigments in urine cannot be destroyed by radicals. This is due to the multiplicity of matter present in urine that react very fast with radicals. Thus, an indirect reaction of ozone with pigments in urine is very unlikely.

While the colour of fresh yellow water disappeared quickly during ozonisation, a subsequent colour increase could be observed in samples treated with small amounts of ozone. This is not re-colouring, as often assumed, but equivalent to the normal colour increase of fresh yellow water, meaning a later transformation of urobilinogen to urobilin. Yellow urobilin differs from the colourless urobilinogen in having a double-single- bond between the two central rings. Because of this central double bond a conjugated $\pi\pi^*$-electron-system can be formed that absorbs the blue fraction of light, which causes the yellow colour. At the same time this central double-bond seems to be the point of attack for a direct reaction with ozone according to the Crigée-mechanism (Figure 11). Urobilinogen, not having this central double-bond will not react with ozone, and is thus still present and able to form new urobilin after treatment with ozone. This leads to a new increase in yellow colour.

After long ozone treatment of fresh yellow water an increase of colour could not be observed. This seems to be also due to low pH, which occurs after long ozonisation. Then the reaction urobilinogen $\rightarrow$ urobilin + $2H^+$ is prevented.

UVC-radiation accelerated the normal colour increase (transformation of urobilinogen to urobilin) of yellow water at first. Later reduction in colour during UVC-radiation may be due to ozone, formed from dissolved oxygen by UVC radiation below 240 nm, and also to photochemical effects, since the discussed pigments absorb the whole spectral UV-bandwidth very strongly.

Biologically pre-treated yellow water had a very constant colour. During this process nearly all urobilinogen is transformed to urobilin. Ozonisation of biologically pre-treated yellow water yielded rapidly to lasting, colourless results, which confirms the theory of urobilinogen transformation to urobilin during biologic pre-treatment. UVC-treatment of biologically pre-treated yellow water also produced decolourising, although the input of electrical energy was about 60 times higher than during ozone treatment.

While the input of electrical energy for ozone treatment of fresh urine was comparably low, only less than 10% of this energy was needed for ozonisation of biologically pre-treated yellow water for reaching a lasting, absolutely colourless liquid. Enhancement of energy efficiency is possible in both decolourising methods.

Outlook

For concepts in which urine is used as process water, while its colour is disturbing this application, a colourless product can be yielded by biological pre-treatment and ozonisation. Also UVC-light (254 nm + 185 nm) will lead to decolourisation at higher energy input.

Components such as micro-pollutants inhibit the direct use of urine as fertiliser. Since the matrix of contained components is very widespread and diverse, tracers are needed to detect unwanted substances. Because of the chemical structure of the colour causing pigments, it is now under investigation to what extend the occurrence of these easy detectable pigments can be used as indicatoro elements in processes for the removal of micro-pollutants.

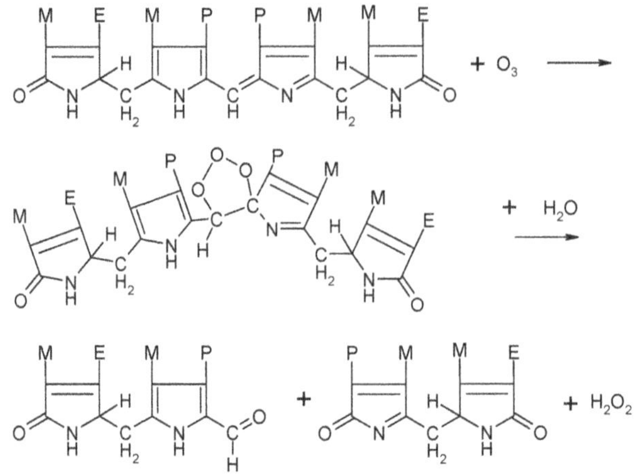

Figure 11 Possible breakdown of urobilin caused by ozone

Additional processes (thermal, component separation e.g. by crystallisation) and also the effect of these and the above mentioned processes on the removal of colour and micro-pollutants, as well as the correlation of colour removal and destruction of micro-pollutants are now the next step that needs to be investigated.

References

Bünning, G. (1996) Kontinuierliche Wasserdesinfektion mittels Ozon in Rohrreaktoren, DissertationsDruckDarmstadt GmbH, Darmstadt.

CIBA-Geigy (1977) Wissenschaftliche Tabellen Geigy, CIBA-GEIGYAG, Basel.

Gottschalk C., Libra J.A., Saupe A. (2000) Ozonation of Water and Waste Water, Wiley-VCH, Weinheim.

Gulyas, H., Breuer, K., Lindner, B. and Otterpohl, R. (2004) Screening of chemical oxidation processes and other methods for decolorization of urine for it's re-use as toilet-flush liquid in ecological sanitation systems, 3rd International Conference on Oxidation Technologies for Water and Wastewater Treatment - AOP's for Recycling and Reuse, No_57, CUTEC Institut GmbH, Clausthal-Zellerfeld.

Gulyas, H. (2000) Verfahren zur Entfernung organischer Abwasserinhaltsstoffe, aus Regenwasserbewirtschaftung/ Bauleitplanung, Eigenüberwachung, dezentrale Abwasserbehandlung, Geruchsbekämpfung, Privatisierung / 12. Kolloquium und Fortbildungskurs zur Abwasserwirtschaft, GFEU e.V., TUHH, Hamburg.

Jönsson, H. (2002) Urine separating sewage systems – environmental effects and resource usage, Water Science and Technology, Vol. 46 No. 6-7 pp 333–340.

Lange, J. and Otterpohl, R. (2000) Abwasser, Handbuch zu einer zukunftsfähigen Wasserwirtschaft, MALLBETON GmbH, Donaueschingen-Pfohren.

Lindner, B., Tettenborn, F., Gulyas, H., Breuer, K. and Otterpohl, R. (2003). Saving water and recovering nutrients by a closed loop system for toilet flushing - a pre test, Proceedings of the 2nd International Symposium on Eco-logical Sanitation, Lübeck, Deutsche Gesellschaft für Technische Zusammenarbeit GmbH, Eschborn.

Vinneras, B., Jönsson, H., Salomon, E. and Stintzing, A.R. (2004) Tentative guidelines for agricultural use of urine and faeces, Proceedings of the 2nd International Symposium on Ecological Sanitation, Lübeck, Deutsche Gesellschaft für Technische Zusammenarbeit GmbH, Eschborn.

Bacterial water quality of rainwater fed domestic hotwater systems

A.T. Spinks*, R.H. Dunstan, P. Coombes*** and G. Kuczera******

*School of Environmental & Life Sciences, University of Newcastle, Callaghan, NSW, 2308, Australia. (E-mail anthony.spinks@studentmail.newcastle.edu.au)
**School of Environmental & Life Sciences, University of Newcastle, Callaghan, NSW, 2308, Australia. (E-mail birhd@cc.newcastle.edu.au)
***School of Environmental & Life Sciences, University of Newcastle, Callaghan, NSW, 2308, Australia. (E-mail pcoombes@mail.newcastle.edu.au)
****School of Civil, Surveying, and Environmental Engineering, University of Newcastle, Callaghan, NSW, 2308, Australia. (E-mail george.kuczera@newcastle.edu.au)

Abstract Rainwater harvesting within the urban environment is being increasingly recognised for its important role in reducing pressures on mains water supply and reducing stormwater runoff. The extent to which rainwater tanks can improve our management of the urban water cycle depends on the level of usage of the rain-harvested water. By incorporating the hotwater service into the suite of uses of harvested rainwater, substantial systemic benefits can be realised. This paper discusses some of the water quality implications of using harvested rainwater through hotwater systems. Field sampling of hot and cold water taps supplied by rainwater tanks was conducted over a 12 month period, together with laboratory conducted thermal destruction experiments on enriched rainwater samples. The field results showed substantial reductions (85–100%) of *Escherichia coli*, total coliform, *Pseudomonas*, and Heterotrophic Plate Count concentrations in harvested rainwater passing through hotwater systems maintained between 50–70 °C. The results of the laboratory thermal destruction experiments indicated that the time required to reduce a heterogeneous bacterial population at 55°C over each of the first three log reductions were 10, 190 and 400 seconds, respectively. The hygienic quality of harvested rainwater was greatly improved after passing through hotwater systems maintained at adequately high temperatures.

Introduction

The sustainability of the urban water cycle has come under increasing pressure over the last century. Freshwater supplies, including surface waters, rivers, and groundwater reservoirs, face increasing pressures from growing urban populations demanding increasing rates of water consumption. Environmental degradation of receiving waterways such as creeks and lakes caused by increasing volumes of poor quality stormwater runoff have resulted from the expansion of impervious surface area. However, the practice of rainwater harvesting within the urban environment has the potential for substantial mitigation of these pressures. The installation of rainwater tanks in new developments and retrofitting existing houses may result in both environmental and economic savings (Coombes *et al.* 2000). The extent of benefits provided by rainwater tanks is determined by the extent to which the harvested rainwater is used. The use of rainwater for hotwater applications has been shown to be a significant factor in maximising the potential impact of rainwater tanks on urban water cycle sustainability (Coombes *et al.* 2002). The constant drawing down of the rainwater tank through hotwater use provides larger available tank capacity for rainwater harvesting in subsequent rain events and provides substantial reductions to mains water demand and peak demand.

However, heated rainwater must be of sufficiently high quality before the widespread recommendation to incorporate rainwater into hotwater systems can be made. Heating water as a means of disinfection is an ancient practice. Widespread recommendations to boil water are still given by water suppliers during times of potential contamination. However, domestic hotwater systems operate at sub-boiling temperatures, and the acceptability of the quality of hotwater supplied from rainwater tanks is under debate as only limited published data is available. Concern has arisen from a number of outbreaks of Legionellosis, caused by inhalation of *Legionella pneumophila*, reportedly from contaminated hotwater systems supplied by community water supplies. The aims of this study were, firstly, to investigate the water quality in five urban rainwater tanks and associated hotwater systems operating at a variety of temperatures, and secondly, to examine the thermal resistance capacities of heterogeneous populations of bacteria found in rainwater tanks.

Methods

Rainwater tanks and hotwater samples

Cold and hot water samples were collected from rainwater tanks from five houses within the Hunter Region/Central Coast areas of New South Wales, Australia. System 1 was comprised of two joining concrete tanks with combined capacity of 35 kL located in a low density semi-urban suburb. The system included a drown pipe, which remained submersed at all times, allowing collection of rainwater from all roof surfaces. No screens or meshes were installed to any of the tank inlets. The tank water was used for all domestic purposes and no connection to mains water was available at the site. Site 2 was a 9 kL polyethylene tank, including a drown pipe, located in an urban area in close proximity to major roads. The tank water was used for all domestic purposes with mains water trickle top-up occurring when tank capacity fell below approximately 10%. Sites 3 and 4 were located in a leafy urban area near a major bush reserve. Site 3 included a 9 kL galvanised iron tank with first flush device. Tank water was used for all household purposes and was topped-up with mains water during low capacity. Site 4 was a 5.4 kL galvanised iron tank used for outdoor, toilet, and hotwater uses. A leaf diverter was used but no first flush device was installed, with trickle top-up from mains water. Site 5 comprised two Aquaplate tanks joined to give combined capacity of 9.06 kL, located in close proximity to a heavily industrial area. The tank water was used for outdoor, toilet, and hotwater purposes with trickle top-up from mains water.

Bacteriological samples were collected in sterile 500 mL polyethylene jars and analysed within 6 hours of collection. The polyethylene jars with hotwater samples were placed in cold water immediately upon collection and kept in the dark during transport. Samples were analysed using Membrane Filtration (Millipore) with enumeration of *Escherichia. coli* and total coliform bacteria on m-ColiBlue24® broth (Prod.# M00PMCB24) by filtering 100mL and 10 mL, *Pseudomonas* spp. with Pseudomonas Selective broth (Prod.# M00000P2P) by filtering 100 mL and 1 mL, and Heterotrophic Plate Counts (HPC) on m-Heterotrophic Plate Count Medium (Prod.# M00000P2S) by filtering 1 mL and 0.1 mL.

Thermal destruction experiments

Laboratory experiments were conducted to examine inactivation rates of a heterogeneous bacterial population at 55 °C. A 200 mL cold water sample was taken from Tank 2 and enriched for 48 h followed by thermal destruction experiments. Enrichment was necessary to produce bacterial concentrations high enough to allow thermal destruction characteristics to be assessed over several log reductions. The enrichment step involved splitting the sample into two 100 mL portions with the addition of 100 mL Brain Heart Infusion (BHI – Oxoid) broth followed by incubation of the split portions at 24 °C and 37 °C for 24 h before re-combining the samples and incubating at 24 °C for a further 24 h. This was designed to allow the culturing of aerobic species of both environmental and enteric bacteria into the stationary growth phase, which has previously been demonstrated as the growth phase in which bacteria possess maximal heat resistance. Cells were harvested by centrifugation and re-suspended in 1.5 mL sterile water.

200 mL of sterile deionised water was heated to 55 °C before inoculation with 1mL of sample. A magnetic stirring bar was set to produce a slight vortex to allow rapid mixing, with sampling occurring at pre-determined intervals thereafter. Temperature was recorded using two Resistance Temperature Detectors (RTDs) which logged directly into a Dataker DT50 data logger connected to a laptop computer providing temperature profiles at one-second intervals using the DeTransfer software. Surviving bacteria were enumerated by spread-plating appropriate dilutions onto R2A agar (Oxoid) and incubating at 37 °C for 48 h.

Results

All cold and hotwater samples were taken concurrently on the same day so that a direct assessment of hotwater system performances could be made. Physicochemical properties of the hotwater changed slightly, with increases in pH between 0.5 and 1.5 pH units from the cold water samples which ranged between 5.7 – 8.4. The results show substantial reductions in bacterial concentrations when water is passed through hotwater systems ranging between 50-70°C. In three of the systems *E. coli* reductions were 100%, although these systems generally had low initial concentrations of *E. coli*, as seen in Table 1.

Table 1 Average *E. coli* concentrations in Cold and Hot water samples (100 mL)

System	Number of samples (*N*)	Number of *E. coli* (Cold Water)	Number of *E. coli* (Hotwater)	Hot water Temperature	% Reduction
1	10	0.3	0	66.3 °C	100
2	8	1429	0.1	50.7 °C	99.991
3	8	113	1.3	65.4 °C	98.893
4	4	7	0	60 °C	100
5	5	8	0	69.2 °C	100

Large variations in bacterial concentrations of the harvested rainwater were found in some of the systems. On three occasions Tank 2 had *E. coli* concentrations above 1000 CFU/100mL (maximum 6800 CFU/100 mL) while two samples had less than 10 CFU/100mL and four samples contained less than 60 CFU/100 mL. In total, 17 (49%) cold water samples were negative for *E. coli*. After passing through the associated hotwater systems, 31 (89%) of the samples were negative for *E. coli*. Three of the four positive samples were from the same system, and contained 1, 1, 2, and 7 CFU/100 mL. Interestingly, system 3 was slightly less effective at reducing *E. coli* concentrations than system 2, despite significantly higher temperatures in the hotwater system.

All rainwater tank samples were positive for total coliform bacteria, ranging between 1–26,600 CFU/100 mL. All hotwater systems averaged over 99% inactivation of total coliform bacteria as seen in Table 2, although only 10 (29%) hotwater samples were negative for total coliforms, ranging between 0–90 CFU/100 mL. System 2 contained the highest number of total coliforms in both the cold and hot water samples. *Bacillus* spp. were positively identified within the total coliform counts on the chromogenic mColiBlue24® media, an issue currently under further investigation.

Table 2 Average Total coliform concentrations in Cold and Hot water samples (100mL)

System	Number of samples (*N*)	Number of total coliforms (cold Water)	Number of total coliforms (hot water)	Hot water Temperature	% Reduction
1	10	1417	1.8	66.3 °C	99.873
2	8	7269	29.8	50.7 °C	99.591
3	8	718	4.3	65.4 °C	99.408
4	4	679	3	60 °C	99.558
5	5	787	1.6	69.2 °C	99.797

Pseudomonas spp. appeared highly sensitive to thermal stress and all hotwater systems achieved greater than 99.8% reduction. Hotwater system 2, operating at around 50 °C, achieved reductions in Pseudomonas comparable to the other four systems operating between 60–70 °C. Temperature appeared to be more crucial to the percentage of inactivation of HPC, as seen in Table 4. The residence time in the hotwater system was not measured but may be an important factor. Spore-forming bacteria are known to have significant thermal resistance capacities, and therefore, the profile of the initial bacterial community entering the hotwater system is also likely to be a significant factor in the efficacy of hotwater system performance.

Table 3 Average *Pseudomonas* spp. concentrations in Cold and Hot water samples (100mL)

System	Number of samples (*N*)	Number of *Pseudomonas* (Cold water)	Number of *Pseudomonas* (Hot water)	Hot water Temperature	% Reduction
1	10	11630	0.2	66.3 °C	99.998
2	8	14868	19.6	50.7 °C	99.868
3	8	4306	0.4	65.4 °C	99.991
4	4	5693	4.5	60 °C	99.921
5	5	5773	0.5	69.2 °C	99.991

Table 4 Average HPC concentrations in Cold and Hot water samples (1mL)

System	Number of samples (N)	Number of HPC (Cold water)	Number of HPC (Hot water)	Hot water temperature	% Reduction
1	10	207	15.6	66.3 °C	92.456
2	8	2640	395	50.7 °C	85.038
3	8	396	3.5	65.4 °C	99.117
4	4	483	14.6	60 °C	96.96
5	5	1045	0.3	69.2 °C	99.976

Inactivation of the enriched heterogeneous population at lethal temperature of 55°C for the first log reduction (90%) required 10 seconds. The second log reduction (90–99%) took approximately 190 seconds, the third log reduction (99–99.9%) took an additional 400 seconds, and the fourth log reduction (99.99% inactivation predicted from extrapolation) took approximately 1400 seconds.

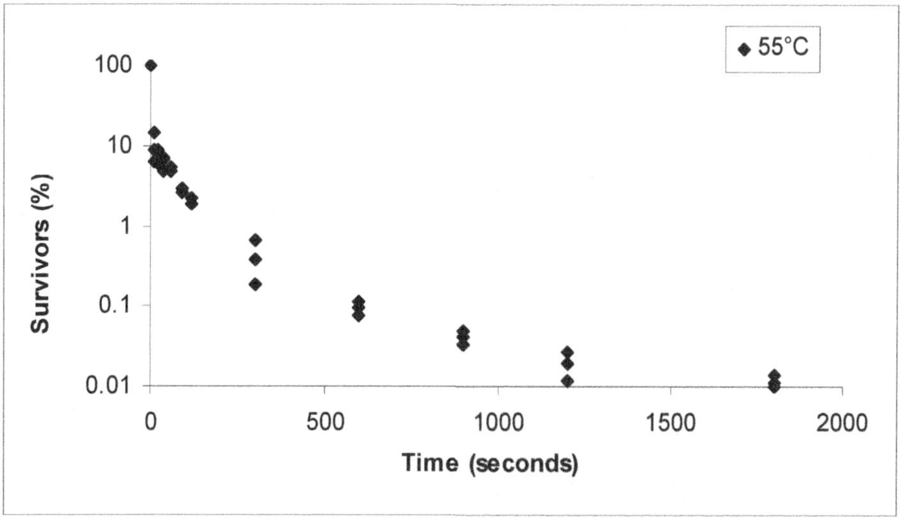

Figure 1 Inactivation of Heterogeneous Bacterial Population at 55°C

Discussion

The quality of stored rainwater varied significantly between systems and over time. System 2 was consistently more contaminated than the other systems with none of eight water samples complying with ADWG for *E. coli*. The presence of a large tree overhanging the roof catchment which was known to house bats may have been a significant contributor to contamination. Colour and odour complaints had previously been made regarding the aesthetic quality of the water during the study period, and felling of the tree post-study appeared to improve water quality. System 1 maintained relatively high quality rainwater with 80% of cold water samples complying with ADWG for *E. coli* despite having employed no water quality protection measures such as first flush devices, leaf diverters, or screens on inlets. Two lizard carcases were also recovered from the tank sludge and several overhanging trees were present, although the owners did not attribute any gastrointestinal complaints to the consumption of the untreated tank water. Four of the five systems relied on mains water to top-up the tanks when capacity fell below approximately 10%. The input of chlorinated mains water may have had an impact on tank water quality although this was not captured in the present study.

The hotwater systems did achieve substantial reductions in the numbers of viable bacteria in the heated rainwater. All hotwater systems monitored in this study were storage hotwater tanks. The timing of sampling within the cycle of the hotwater systems was likely to be an important factor, as longer residence time in hotwater results in greater reductions of bacteria. The level of inactivation was not only associated with the temperature of the hotwater system. The initial composition of bacterial communities within the rainwater tank may also significantly influence the degree of reduction. The results of the thermal destruction experiments indicated that significant levels of reduction are achieved upon initial exposure to heat, presumably due to inactivation of heat-sensitive species, but that the rate of population reduction is significantly slowed after a period, presumably due to the residual presence of spore-forming bacteria. Spore-forming bacteria, such as *Bacillus* and *Clostridium* spp., are known to possess heat resistance capacities beyond that which could reasonably be imparted by domestic hotwater services. However, while some species of spore-forming bacteria are pathogenic, these are rarely transmitted through water and consequently are not addressed as bacterial species of concern in the Australian Drinking Water Guidelines (ADWG).

Previous studies have also noted reductions in bacterial loads of harvested rainwater passing through domestic hotwater systems. In Newcastle, monitoring of the Figtree Place water sensitive urban design project showed that in the electric hotwater systems used on these premises total coliform counts were reduced from an average 112 CFUs/100 mL in the rainwater tanks to 0 CFU/100 mL from the hotwater system tap, while *Pseudomonas* were reduced from 29,500 CFUs/100 mL to 0CFUs/100mL (Coombes *et al.* 2000). Substantial reductions in bacterial loads through instantaneous hotwater systems have also been noted. The results of monitoring a single instantaneous hot water system set to 55°C demonstrated a 4-log reduction in *Pseudomonas* concentrations in less than 1 minute of exposure (Coombes *et al.* 2003).

One of the major threats to public health associated with hotwater systems is that of infection caused by *Legionella pneumophila. L. pneumophila* is the aetiological agent of Legionnaires disease, an acute form of pneumonia, which most commonly infects the respiratory tract of immuno-compromised individuals. *L. pneumophila*-associated infections occur as a result of the inhalation of contaminated aerosols. Ingestion of high concentrations of *L. pneumophila* does not cause harm. A limited amount of research has been conducted which investigated the heat tolerance of *L. pneumophila* in a water medium (Stout *et al.* 1986, Dennis *et al.* 1984). Research findings generally indicated that *L. pneumophila* has little thermal resistance capacity above 60 °C. As a result of this research, Australian standards (AS3500.4.2) have been developed that state that hot water systems should be maintained above 60 °C in order to inhibit the growth of *L. pneumophila*. Lye (1991) found that three hotwater systems supplied by harvested rainwater maintained below 53 °C were positive for *Legionella*-like isolates. The two systems maintained above 60 °C were negative for such isolates. When the operating temperatures of the three hotwater systems operating below 53 °C were increased to above 60 °C, *Legionella*-like isolates could no longer be isolated.

Martinelli et al. (2000) found that *L. pneumophila* were isolated much more frequently from hot water tanks (30%) as opposed to instantaneous hotwater systems (6.4%) in an area of Italy suffering an endemic outbreak of Legionellosis. The study concluded that temperature was critical, as the hotwater tanks in the study had been maintaining water at 50 °C ± 5 °C, while the tap water averaged 45 °C and dropped to 40 °C during flow. Instantaneous systems, however, always maintained temperatures above 60 °C (Martinelli *et al.* 2000). Ezzeddine et al. (1989) found that the main reservoir for *L. pneumophila* were mixing tanks where hotwater at 60–65 °C mixed with cold water to achieve 45°C. Maintaining temperatures of 60 °C or accelerating flow rate were found to be the most effective methods of controlling *L. pneumophila* (Ezzeddine *et al.* 1989). Lee *et al.* (1988) discovered that water temperatures in electrically heated tanks were significantly lower than in gas-heated tanks. They concluded that the presence of *L. pneumophila* was associated with systems maintained below 48.8°C. Wadowsky et al. (1985) detected *L. pneumophila* in water and sediment samples taken from hotwater systems maintained at 30–54 °C, but not from systems maintained at 71–77 °C.

The nature of the contact with water must be understood before an evaluation of risk can be made. The uses of hotwater are typically quite restricted within households, generally limited to showering, washing dishes, and cooking. Few people drink directly from the hotwater tap without prior additional heating such as boiling for coffee/tea or cooking. Ingestion of large amounts of water while showering is not common practice, and needs to be considered when assessing hotwater quality against ADWG which are based on the ingestion of 2 L per day (enHealth 2004). The threat of respiratory infection caused by inhalation of *L. pneumophila* may be of greater significance than that posed by direct ingestion. However, hotwater systems maintained at 60°C in accordance with Australian Standards should pose minimal health risks from either ingestion or inhalation. The immuno- and nutritional-status of a population also influences susceptibility to illness. Immuno-compromised individuals are at greater risk of infection from many sources including harvested rainwater and hotwater systems. It is recommended that immuno-compromised individuals exercise care when using harvested rainwater, such as boiling water intended for consumption.

Conclusions

Large reductions in *E. coli*, Total coliform, *Pseudomonas*, and HPC concentrations were consistently observed in harvested rainwater passing through domestic hotwater systems maintained between 50–70°C. Low numbers of viable total coliforms were present in hotwater samples, although further identification of isolates showed that these were predominantly *Bacillus spp.* This may have significant implications when assessing rainwater-fed hotwater systems as current ADWG operate on a presence/absence criteria for determining the acceptability of a water. Laboratory experiments showed that enriched bacterial populations cultured from harvested rainwater decreased rapidly with initial exposure to thermal stress, with reductions in activation rates during subsequent heating. All hotwater systems operating between 50-70°C performed well achieving substantial reductions in bacterial loads, although due to the possible presence of *L. pneumophila* in hotwater systems, it is recommended that hotwater services be maintained above 60 °C.

Acknowledgements

This work was supported by a grant from the Australia Research Council in conjunction with Hunter Water Corporation, Hunter Councils, Gosford Wyong Councils' Water Authority, Brisbane City Council, Department of Commerce and NSW Public Health Unit.

References

Coombes, P. J., Argue, J. R. and Kuczera, G. (2000) Figtree Place: a case study in water sensitive urban design (WSUD). *Urban Water* **4** (1), 335–343.

Coombes, P. J., Kuczera, G. and Kalma, J. D. (2003) Economic, water quantity and quality impacts from the use of a rainwater tank in the inner city. *Australian Journal of Water Resources* **7** (2), 101–110.

Coombes, P. J., Kuczera, G., Kalma, J. D. and Argue, J. R. (2002) An evaluation of the benefits of source control measures at the regional scale. *Urban Water* **4** (4), 307–320.

Dennis, P. J., Green, D. and Jones, B. P. 1984 A note on the temperature tolerance of *Legionella*. *Journal of Applied Bacteriology* **56**, 349–350.

enHealth (2004) *Guidance on use of rainwater tanks*. Australian Government, National Public Health Partnership, Canberra.

Ezzeddine, H., Van Ossel, C., Delmee, M. and Wauters, G. (1989) *Legionella* spp. in a hospital hot water system: effect of control measures. *Journal of Hospital Infection* **13** (2), 121–131.

Lee, T. C., Stout, J. E. and Yu, V. L. (1988) Factors predisposing to *Legionella pneumophila* colonization in residential water systems. *Archives of Environmental Health* **43** (1), 59–62.

Lye, D. J. (1991) In *Proceedings of the 5th International Concference on Rainwater Catchment Systems*Keelung, Taiwan, pp. 80–86.

Martinelli, F., Caruso, A., Moschini, L., Turano, A., Scarcella, C. and Speziani, F. (2000). A comparison of *Legionella pneumophila* occurrence in hot water tanks and instantaneous devices in domestic, nosocomial, and community environments. *Current Microbiology* **41**, 374–376.

Stout, J. E., Best, M. G. and Yu, V. L. (1986) Susceptibility of members of the family Legionellaceae to thermal stress: implications for heat eradication methods in water distribution sytems. *Applied and Environmental Microbiology* **52** (2), 396–399.

Wadowsky, R. M., Wolford, R., McNamara, A. M. and Yee, R. B. 1985 Effect of temperature, ph, and oxygen level on the multiplication of naturally occurring *Legionella pneumophila* in potable water. *Applied and Environmental Microbiology* **49** (5), 1197–1205.

Microbial risk assessment tool to aid in the selection of sustainable urban water systems

N.J. Ashbolt*, S.R. Petterson*, D.J. Roser*, T. Westrell, J. Ottoson**, C. Schönning** and T.A. Stenström****

*Centre for Water & Waste Technology, School of Civil & Environmental Engineering, University of New South Wales, Sydney, NSW 2052, Australia
**Swedish Institute for Infectious Disease Control, Solna, SE-171 82, Sweden

Abstract Traditionally, the hygienic impact of urban water systems has been assessed by the occurrence of thermotolerant coliforms/*E. coli* at points of exposure or treatment. When some form of treatment is used, pathogens may be present in the absence of these faecal indicators. Quantitative microbial risk assessment (QMRA) is now the preferred approach to assess pathogen risks at points of exposure, based on estimation of pathogen exposure and human dose–response models. The development of a system's life-time approach to applying QMRA is described here (the MRA tool), where different urban water system options can be compared on the basis of infections per month (to account for outbreak potential), the variation/uncertainty in this infection risk (system robustness) and identification of the critical pathogen(s) and pathway(s) of infection (to aid risk management). Overall the MRA tool provides hygiene information appropriate for integration into a systems analysis that may include environmental, social, technical and economic aspects to better assess the sustainability of urban water systems over their intended life-times.

Keywords Pathogens, systems analysis, disease, microbial risk assessment

Introduction

Internationally and within Australia, it is becoming less acceptable to discharge wastewaters to receiving waters, if not due to eutrophication or pathogen problems, simply due to water scarcity (Rathjen *et al.*, 2003). Indeed, based on current sources and likely demands, there is no Australian capital city that can be said to be water secure by 2025. Internationally, there is an urgent need to examine alternative urban water strategies (Bartram *et al.*, 2005). Tools to assess potential human health risks that can be integrated with ecological footprints, triple bottom line reporting and other 'scorecard' approaches to assessing sustainability are wanted. Reliance on faecal indicator bacteria is simply inadequate to cover the range of treatment-resistant pathogens potentially released from modern urban water systems (Ashbolt *et al.*, 2001).

Interestingly, pathogen risks appear to be largely ignored in so-called water-sensitive urban designs of ecological sanitation (EcoSan) alternatives, such as described by Michell *et al.* (2003) and van der Vleuten-Balkema *et al.* (2004) respectively, although the need for their inclusion has been noted (Milburn *et al.*, 2002). This is surprising given the interest for EcoSan options in developing regions and that some 3.1% of global deaths (1.7 million) or 3.7% of disability adjusted life years (DALYs) (54.2 million) are attributable to pathogens via unsafe water, sanitation and hygiene (99.8% of deaths associated with this risk factor are in developing countries, and 90% are associated with deaths of children) (WHO, 2003a).

In contrast, international guidelines for waters (WHO, 2003b; WHO, 2004) and current revisions of various Australian guidelines (NHMRC, 2004) strongly advocate an assessment of the performance of barriers to pathogens from sources through the water supply system to recipients. This approach is now the preferred way to manage all water risks, and the water industry places pathogens as the highest priority hazard. Nonetheless, suitable tools are not yet generally available to support agencies in assessing pathogen risks along with environmental, social and technical aspects in the context of a systems analysis for sustainability assessment.

This paper describes the development of a microbial risk assessment (MRA) tool, that allows integration with environmental, economic, socio-cultural and technical function in association with the Swedish Sustainable Urban Water Project (MISTRA, http://www.urbanwater.org/default_eng.htm), and builds on what has been previously published by the consortium (Malmqvist *et al.*, 2000). In taking a systems approach the MRA tool attempts to account for system failures and intrinsic variability of unit operations over their intended life-times, and in theory can be applied to a broad range of urban water systems.

Materials and Methods

The MRA tool is based on information collected from various small communities within Sweden (Gebers, Vibyåsen & Hammarby Sjöstad in Stockholm, and Vasastan in Gothenburg). MRA data for Hammarby Sjöstad are presented here to illustrate two urban water system structures; a) conventional 'centralised system', and b) blackwater-yellow water (urine) 'source-separation system', based on the life-cycle pathogen flux analysis for index pathogens described by Fane and Ashbolt (2000).

System Structures

The drinking water supply system for Hammarby Sjöstad is proposed to be the same for both the 'Centralised' and 'Source-separation' cases, so there was no relative difference in pathogen risks between the two cases for drinking water exposures. Therefore, attention was paid to the wastewaters, where for the centralised case, wastewater was conveyed by sewer to one large activated sludge plant (with screening, grit chamber, pre-aeration, pre-sedimentation with coagulant, activated-sludge with nitrification-denitrification, and coagulation-sand filtration) with no disinfection of effluent and sludge processed by pasteurization, anaerobic thermophilic digestion, and the KREPRO (Kemwater, Helsingborg, Sweden) nutrient recovery process for soil application of biosolids. A flow diagram of the 'Source-separation' system is provided in Figure 1, with the main difference being that urine was separated by urine-diversion toilets, stored (for disinfection) and used as a fertilizer for crop production.

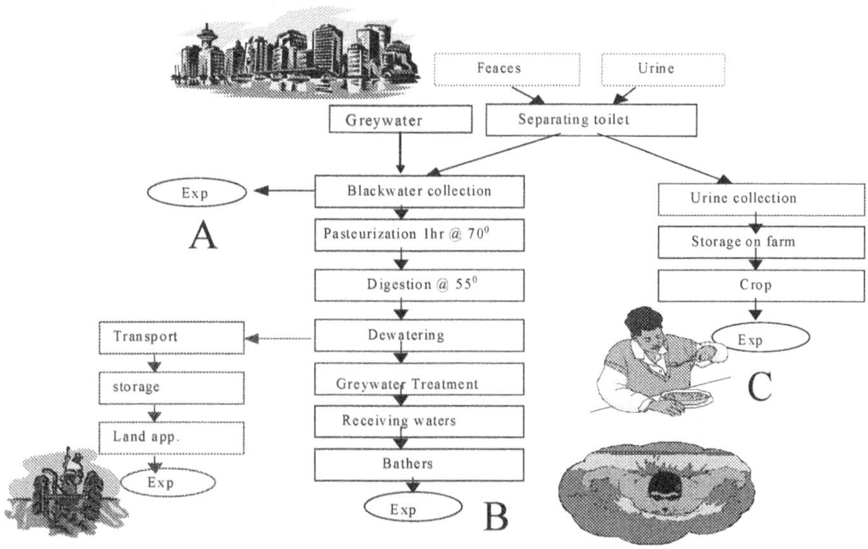

Figure 1 'Source-separation' case for Hammarby Sjöstad. (Drinking water provisions same as for centralised case.) Possible exposure points of concern identified: A - workers fixing plumbing & sewage overflow; B - recreational water users in receiving water; and C - consumers of crops grown with urine fertiliser.

Data sources

Epidemiologic and literature data along with direct measurements in the Swedish systems were utilised to estimate viral, bacterial, and parasitic protozoan pathogen ranges in source materials (faeces, urine and sewage). Pathogen removal by the key system units (described above) were modelled by reference pathogen numbers (rotavirus, *Campylobacter, Cryptosporidium* and *Giardia*). Reference pathogen numbers in source faeces/urine/sewage were described as probability density functions (PDFs) (Table 1) and their dilution/removal/inactivation described as the pathogens 'flowed' through each system structure. Overall, helminth risks were considered low, but were included from estimates in sewage (with *Ascaris lumbricoides* as reference).

Table 1 Epidemiologic data used to estimate pathogen probability density functions (PDFs) in sewage (modified from Höglund *et al.*, 2002b)

Parameter	Enteric virus (Rotavirus)	Bacterium (*C. jejuni*)	Protozoan (*C. parvum*)	Protozoan (*G. lamblia*)
Yearly incidence of infection (%)	0.95	15.6	0.31	0.84
Disease rate if infected (%)	75[a]	23	39	50
Log_{10} Excretion time (days)	N*(1.0,0.30)	N(1.18, 0.325)	N(1.48,0.173)	N(1.18,0.325)
Log_{10} (Excretion density) (no/g faeces)	N (10, 1)	N (8, 1)	N (7, 1)	N (7, 1)
Estimated # in sewage**				
Mean (per L)	1.53×10^7	3.38×10^6	1.27×10^4	2.13×10^4
50th percentile (per L)	8.97×10^5	1.87×10^5	8.97×10^2	1.24×10^3
95th percentile (per L)	4.55×10^7	1.03×10^7	4.13×10^4	6.29×10^4

*Normal distribution defined by parameters (mean, standard deviation)

**Assuming 150 g faeces/person.day and 135 litres per capita.day water released to sewer.

[a] Only for children and the aged (assumed to be 15% of population)

Concern that our 'epidemiologic' approach (Table 1) may over estimate pathogens in faeces/sewage was examined with sewage datasets from Australia (taking into account minor differences in reported disease prevalence's). Overall there was good agreement (within an order of magnitude) for the reference pathogens, but higher variability exhibited by the epidemiologic approach than assayed (from data collected over one year) (Figure 2). Virus numbers in sewage, however, were estimated to be some three order of magnitude higher than numbers cultured (infective) in sewage, which may reflect reality (virus recovery was not reported, but could be 10-40%). Furthermore, non-culturable viruses such as Norovirus, have been reported by non-culture (i.e. PCR methods) at levels similar to those reported in Table 1 (Lodder *et al.*, 1999). Therefore both virus estimations were used in the MRA tool to provide a feel for the uncertainty in the overall estimation of viral infection risks.

MRA tool structure

The MRA tool is still under development, but is designed to allow user selection of system components, which defines the specific datasets for pathogen system performance (removal/inactivation) and reliability, coded in the software package Analytica® (Version 3, Lumina Decision Systems Inc., Los Gatos, California). Each unit operation in the system structure has pre-described variability and pathogen performance data (e.g. for sand filter, chlorination, distribution etc., called a Hygiene Modifying Process [HMP]) (Figure 3), and users confirm likely points of exposures (e.g. in home, recreation water site, drinking water etc.). Dose-response models for the reference pathogens were used as described by Haas *et al.* (1999) using relevant exponential and beta-Poison models, with the maximum risk curve ($r = 1$ in the exponential model) (Teunis and Havelaar, 2000) applied for the helminth *Ascaris lumbricoides*, should it be considered important (e.g. for sludge application).

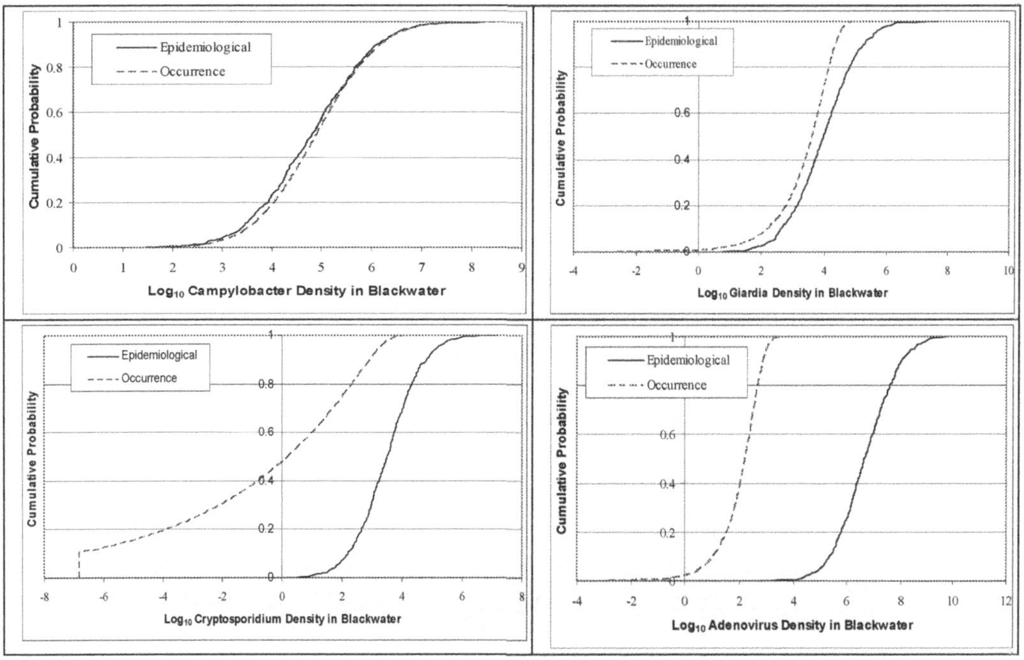

Figure 2 Example output from MRA tool illustrating spread in the pathogen PDF's when estimating numbers in sewage from epidemiologic (solid line) (see Table 1) or occurrence (dashed line) datasets

Analytica® provides influence diagrams which are hierarchical, providing the user with an intuitive means to see increasing detail, as illustrated in Figures 4 & 5. Estimated infections for the life-time of a system structure were compared on a monthly basis by running Monte Carlo simulations of pathogen numbers through each system thousands of times, so as to capture acute effects which may lead to outbreaks. System robustness was estimated based on the range in variability/uncertainty of annual infections, which also helps to identify critical pathogens and pathways of infection, and therefore where to target management actions. Hence, relative comparisons between systems are the prime output of the model, although disease risks are also expressed in a qualitative manner based on the conditional probability of infection (Table 2), to enable integration with the outputs from other sustainability assessments.

In our previous quantitative microbial risk assessments of Swedish urban water system (Höglund *et al.*, 2002b; Ottoson and Stenström, 2003; Westrell *et al.*, 2003; Westrell *et al.*, 2004), a large range of sites of potential exposure to pathogens were investigated. The MRA tool makes use of this knowledge and limits potential exposure sites to only those considered more likely and important (although user changeable). For example, sites A-C in Figure 1 and not at the point of biosolids application, given the previous pasteurisation and thermophilic digestion of the sludge.

Results and Discussion

The pathogen group of highest concern varied by point of exposure, but based on overall risk, enteric viruses generally dictated the outcome in both the centralised and source-separated alternatives at Hammarby Sjöstad (and in other system structures studied to date [Höglund *et al.*, 2002b; Ottoson and Stenström, 2003; Westrell *et al.*, 2003; Westrell *et al.*, 2004]). The highest likelihoods of infection resulted from obvious risks, such as exposures to raw sewage during mains breaks/leaks in the centralised case or unblocking urine-diversion pipes in the other; yet few individuals would be exposed during such scenarios. Overall community risks were ranked 'minor' for both systems, but the 'source-separate system' yielded the lower risks, due to less impact from recreation swimming.

Table 2 Risk rankings based on estimated disease rates for each system (from Westrell *et al.*, 2004)

Item	Definition
Catastrophic	Major increase in diarrhoeal disease >25% or >5% increase in more severe disease or large community outbreak (100 cases) or death
Major	Increase in more severe diseases (0.1-5%) or large increase in diarrhoeal disease (5-<25%)
Moderate	Increase in diarrhoeal disease (1-<5%)
Minor	Slight increase in diarrhoeal diseases (0.1-<1%)
Insignificant	No increase in disease incidence (<0.1%)

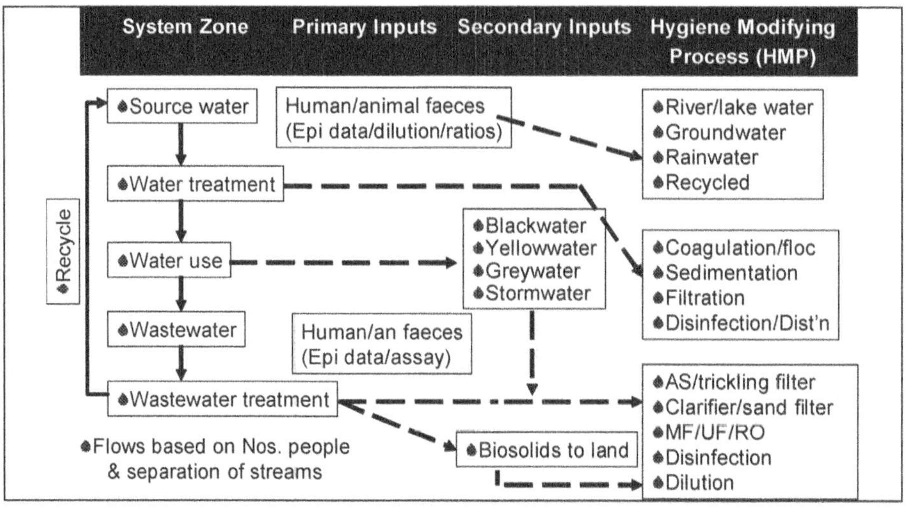

Figure 3 General schematic of the MRA tool for human health assessment

The MRA tool also provides valuable information for risk management. Rankings of infection risk potential by exposure sites enables the identification of control points for pathogen management, within the Hazard Analysis Critical Control Points (HACCP) risk management approach being adopted by WHO and other water guideline setting agencies. The ranking of microbial risks per system also provides the input values for integration with the other measures of sustainability. Given the potential significance of viral infections, control of aerosols becomes an important aspect to workers near raw wastewater streams. Another important control of viruses is sufficient storage of urine before crop application (from one month for cereals to six months for vegetables) at temperatures of about 20°C (Höglund *et al.*, 2002a). Disinfection of wastewater discharged to receiving waters would also have improve the performance of the 'centralised system', then leaving maintenance exposures as the main residual issue for both systems (Figure 6).

As a tool to aid in selecting treatment options, alternative systems can be compared based on relative levels of infection risk, acknowledging the uncertainties in each. Hence, existing and hypothetical systems can be compared, and when integrated with other assessments (environmental, economic, social and technological) can provide management information (e.g. 'scorecard' for each option). During a multi-criteria decision aided (MCDA) process (van Moeffaert, 2003), the MRA tool output can further inform the various stakeholders involved. In such multi-criteria processes, qualitative rankings are all that are generally needed.

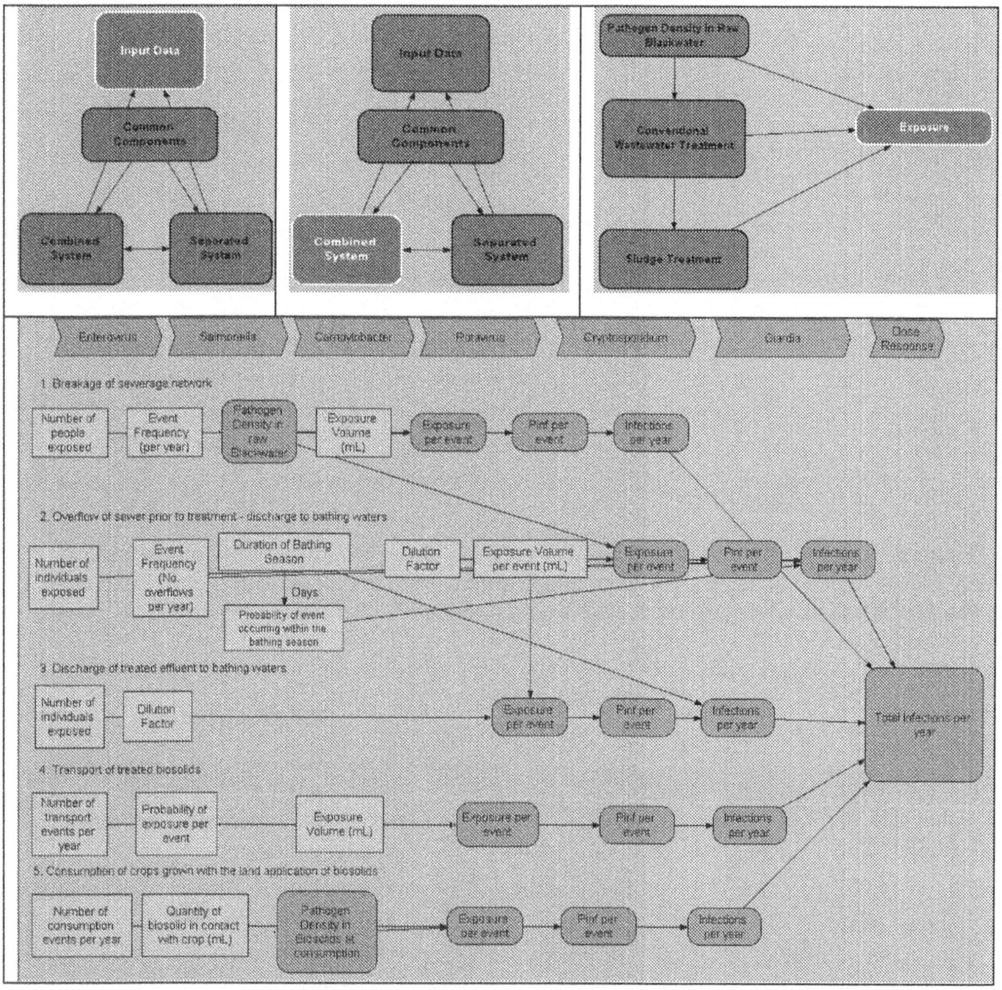

Figure 4 Example influence diagrams within the MRA Tool: all the *Input Data* (top left) within this influence diagram is illustrated in Figure 5; *Combined System* highlighted (top centre) and when the user double clicks it the figure on the top right is reveled; and similarly within the *Exposure* influence diagram, information (bottom diagram) for the five exposure points and range of reference pathogens assessed in the 'combined system' to estimate annual infections.

One of the advantages of using Analytica® is that there is a freely available version of the software to run constructed models, such as the MRA tool. It is envisaged that free versions of the tool will be made available in the medium term, but beta testing of specific systems is currently available through discussion with the authors.

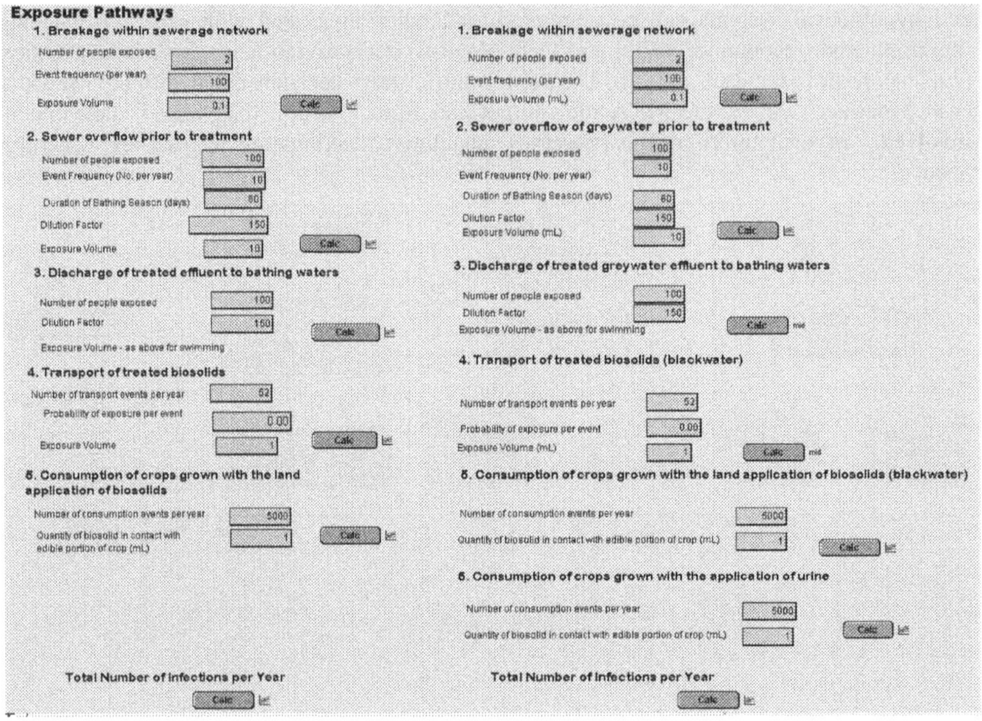

Figure 5 Example influence diagram within the *Input Data* (influence diagram illustrated top left, Figure 4) used to assess the 'Combined System' (left) and separated (right). Each input has default values which can be user-changed and the effect shown (within the 'Calc' boxes)

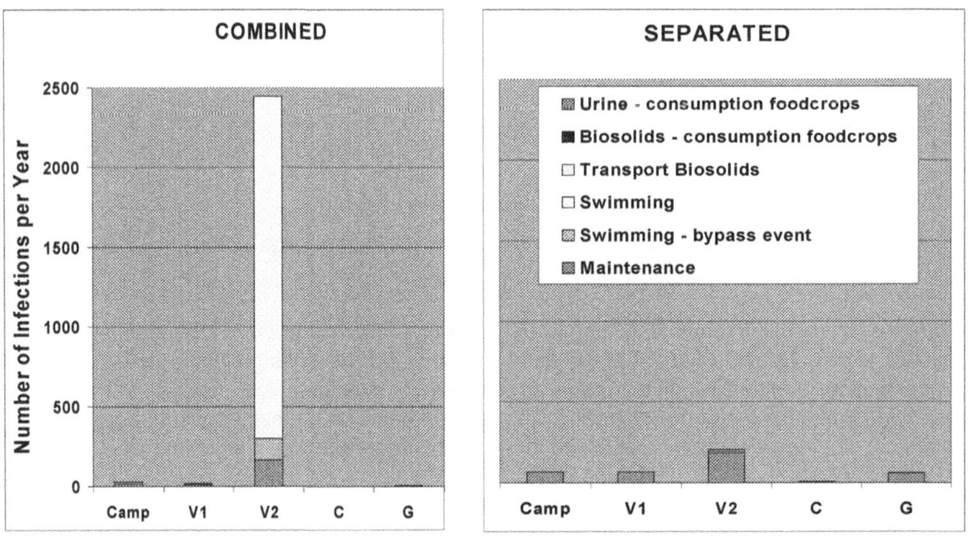

Figure 6 Estimated annual infections for 'Combined' (left) and 'Separated' (right) system structures for reference pathogens (camp – *Campylobacter*, V1 – & V2 – epi and occurrence virus estimates from sewage. C – *Cryptosporidium*, G – *Giardia*)

Acknowledgements

We thank the various members of the Urban Water Programme who have contributed to discussions in the development of the MRA tool, and in particular Erik Kärrman, Daniel Hellström, Håkan Jönsson and Per-Arne Malmqvist; as well as initial work undertaken with Simon Fane. This work was financially supported by the Swedish Foundation for Strategic Environmental Research (MISTRA) and carried out within the Swedish research programme "Sustainable Urban Water Management" (Malmqvist, 1999).

References

Ashbolt, N. J., Grabow, W. O. K. and Snozzi, M. (2001) *In* Water Quality: Guidelines, Standards and Health. Risk assessment and management for water-related infectious disease, Chapter 13 (L. Fewtrell and J. Bartram, eds), pp. 289–315. IWA Publishing, London.

Bartram, J., Lewis, K., Lenton, R. and Wright, A. (2005) *Lancet* **365**, 810–812.

Fane, S. and Ashbolt, N. J. (2000) *In* Water Recycling Australia (P. J. Dillon, ed.), pp. 11–16. CSIRO Land and Water, Australian Water Association, Adelaide.

Haas, C. N., Rose, J. B. and Gerba, C. P. (1999) Quantitative Microbial Risk Assessment. John Wiley & Sons, Inc, New York.

Höglund, C., Ashbolt, N. J., Stenström, T. A. and Svensson, L. (2002a) *Advances in Environmental Research* **6**, 265–275.

Höglund, C., Stenström, T. A. and Ashbolt, N. (2002b) *Waste Management and Research* **20**, 150–161.

Jakubowski, W., Sykora, J. L., Sorber, C. A., Casson, L. W. and Gavaghan, P. D. (1991) *Water Science and Technology* **24**, 173–178.

Lodder, W. J., Vinj, J., van de Heide, R., de Roda Husman, A. M., Leenen, E. J. T. M. and Koopmans, M. P. G. (1999) *Applied and Environmental Microbiology* **65**, 5624–5627.

Malmqvist, P.-A. (1999) *Vatten* **55**, 7–17.

Malmqvist, P.-A., Ashbolt, N. J., Fane, S., Hellström, D., Jeppsson, U. and Söderberg, H. (2000) *In* Second International Conference on Decision Making in Urban and Civil Engineering (J.-C. Mangin and M. Miramond, eds), vol. 1, pp. 1–13. CUST Clermond-Ferrand, LIP6 Paris and Universit, de Valenciennes, Lyon, France.

Milburn, A., Matsui, S. and Malmqvist, P. A. (2002) *Water Science and Technology* **45**, 195–198.

Mitchell, G., Gray, S., Shipton, B., Woolley, R., Erbacher, J., Egerton, G. and McKnoulty, J. (2003) *Water Science and Technology* **47**, 1–9.

NHMRC. (2004) Australian Drinking Water Guidelines (ADWG) National Health and Medical Research Council, Canberra. http://www.nhmrc.gov.au/publications/synopses/eh19syn.htm

Ottosson, T. A. and Stenström, T. A. (2003) *Water Research* **37**, 645–655.

Rathjen, D., Cullen, P., Ashbolt, N., Cunliffe, D., Langford, J., Listowski, A., McKay, J., Priestley, T. and Radcliffe, J. (2003) Recycling Water for Our Cities. Report to Prime Minister's Science, Engineering and Innovation Council (PMSEIC), 28th November 2003. Federal Government of Australia, Canberra.

Teunis, P. F. M. and Havelaar, A. (2000) *Risk Analysis* **20**, 511–518.

van der Vleuten-Balkema, A. J., Otterpohl, R. and Preisig, H. A. (2004) *In* Proc. 4th World Water Congress, 19–24 of September 2004, Marrakech, pp. CD-ROM. International Water Association, London.

van Moeffaert, D. (2003) MultiCriteria Decision Aid in Sustainable Urban Water Management. MSc. thesis. TRITA-KET-IM 2002:26. Dept. of Chemical Engineering and Technology, Royal Institute of Technology, ISSN 1402-7615, Stockholm.

Westrell, T., Bergstedt, O., Stenström, T. A. and Ashbolt, N. J. (2003) *International Journal of Environmental Health Research* **13**, 181–197.

Westrell, T., Schönning, C., Stenström, T. A. and Ashbolt, N. J. (2004) *Water Science and Technology* **50**, 23–30.

WHO (2003a) *In* The World Health Report 2002. World Health Organization, Geneva.

WHO (2003b) Guidelines for Safe Recreational Water Environments. Vol. 1: Coastal and Fresh Waters. World Health Organization, Geneva.

WHO (2004) Guidelines for Drinking-water Quality Third Edition. Volume 1. World Health Organization, Geneva.

Governance for sustainability in water resource management: the role of collaborative decision making

J. O'Riordan

Adjunct Professor at the University of British Columbia and the University of Northern British Columbia. Former Deputy Minister of the Ministry of Sustainable Resource Management, Province of British Columbia, Canada.

Abstract Sustainability in this paper is defined as the integration of environmental, economic and social values. Integration of these values rather than trade-offs between values requires a different form of governance than has traditionally been provided solely by the public sector. Sustainable decisions require greater collaboration among four institutions that set the framework for modern governance, namely the public sector (various level of elected government), the corporate sector, the civic sector (including not-for-profit groups) and the knowledge sector(universities and colleges). This definition of sustainability embraces eco-efficiency, which entails significant reductions in both water and energy resource use and in waste generation. Eco-efficiency must be encouraged by proper signals such as full cost accounting, voluntary actions by the private sector beyond government regulation, backed up by strict enforcement of legal regulations. The paper briefly reviews a number of case studies in British Columbia, Canada where collaborative decision making has led to more sustainable decisions in water resource management

Keywords Sustainability; collaborative decision-making; state of environment reporting; governance; watershed management; restorative economy; eco-efficiency

Introduction

Traditionally the term "sustainability" is defined as the integration of the three "stools" of the environment, the economy and society into decision-making. This paper proposes that there is a fourth dimension to sustainability – that of governance. The conventional form of governance is the "state centric" model where the institution of elected governments steers society towards publicly defined priorities and goals. In an increasing globalizing world with instant access to vast amounts of information, this model is rapidly giving way to a "society-centered" form of governance where public policy making is formally shared with influential segments of society (Pierre,1999). This paper adopts the second model of collaborative governance as the only viable mechanism to advance sustainability in water resource management and indeed in other socio-economic sectors.

The term "sustainability" is also subject to a range of definitions. For the purpose of this paper it is defined as definite hierarchy of values, namely that stable and functioning ecosystems are essential to support a sustainable and growing economy which in turn can nurture a healthy and just society. There is little doubt that water-based ecosystems are being degraded at a greater pace than the ability of the dominant state-centered governance model can shift public policy to reduce the environmental footprint associated with water use. To turn the corner to a more restorative economy where not only is the rate of resource depletion arrested but in some cases natural capital is restored will require the collective energy of four institutional domains that together provide the framework for governance in a society-centered model.

The first of these domains is the established form of government at the national, regional and local levels. These institutions have the unique powers to pass laws and regulations and establish public policies that influence the other domains. Recently some of these government bodies have embraced the concept of eco-efficiency, in which there is a significant reduction in resource and energy use, increased application of re-use and re-cycling principles, elimination of toxic waste and a shift to a service focus rather than a commodity focus to product use (Willard, 2002). A noteworthy example is the recently released white paper by the State of Victoria government entitled "Our Future Our Water" that uses a wide range of market and pricing mechanisms backed up by regulations to achieve an eco-efficient management of the State's water resources.

The second domain is the corporate sector represented by large to small private companies that deliver a range of goods and services. Recently this sector has begun to embrace the concept of "corporate social responsibility" and "triple bottom line reporting" plus independent certification audits of environmental management practices, which have initiated a transformation to more sustainable resource use (Hawken et al, 2000).

The civic/ not-for-profit sector represents the third domain in this governance framework. It has influenced sustainability through targeted market campaigns to encourage companies to trade only in sustainable products and services. This sector is in a state of flux between the traditional advocacy approach that uses direct action to influence public policy and a more collaborative model that works in partnership with some or all of the other governance sectors to advance sustainability.

The fourth institution is the knowledge sector that provides independent and peer reviewed science to inform decision-makers. To embrace sustainability that inherently requires multi- disciplinary and multi- dimensional research, the university governance model will have to be adapted from one that encourages individual effort in narrowly defined disciplines to one that encourages research across faculties and disciplines in partnership with the government and private sectors.

The examples used in this paper to illustrate the influence of collaborative governance model are based on the author's direct participation and experience. They are by no means comprehensive, nor are they transformational in a substantive sense. True sustainability in water and other resource management will require a much more profound shift in the paradigm that is represented largely by the eco-efficiency approach. It will have to embrace changing the entire system design towards the holistic ecological model if the projected growth in global population and economies is to sustainably supported (Frankel, 1998). In the analogy of making popcorn, first there is a single pop, followed by more pops until the whole mass of popcorn explodes. These single pops are the leaders in collaborative governance that will initiate the transition to sustainability and are briefly reviewed in the balance of this paper.

Water use planning

The British Columbia provincial government, the federal government and BC Hydro, the province's Crown Corporation responsible for electric power generation and transmission undertook a collaborative approach to water use planning for most of BC Hydro's 34 hydro-electric generating facilities. This approach exemplifies many of the principles of society-centered governance and restoration of natural capital outlined in the Introduction

BC Hydro has built over 11,000 MW of generating capacity over the past 70 years. Approximately 90 percent of this capacity is hydro-electric of which a substantial portion was built before 1970 when environmental assessment processes and policies began to influence the design of such facilities. In the 1990's Hydro faced three challenges to its social license to operate in BC. First, the Department of Fisheries and Oceans, Canada (DFO), which has a constitutional responsibility to protect fish and fish habitat under the Federal Fisheries Act (Sect. 35) took legal action against BC Hydro in 1993 on one of its storage reservoirs. Second BC Hydro faced public scrutiny when it completely drained another one of its storage reservoirs for dam maintenance purposes leaving graphic pictures of stranded fish in the middle of a provincial election campaign in May 1996. The government immediately ordered an inquiry led by a prominent environmental lawyer, who found that in practical terms BC Hydro could not operate its reservoirs without harming fish and fish habitat and proposed a collaborative model rather than litigation to resolve these conflicts . The third challenge came under the North American Trade Agreement under which there is a side agreement for environmental cooperation between the three countries. British Columbia was challenged under this agreement that there was an unfair subsidy to electrical generation because the full costs of environmental damage associated with power generation were not reflected in the price charged to trading partners.

As a result of the initial DFO litigation, the provincial government in 1993 directed BC Hydro to undertake an "electric systems operating review" of all its storages to develop "management strategies that would balance power production, flood control, social, environmental and recreational impacts." This directive was accompanied by a pilot project on the Stave-Allouette system located about 50 km east of Vancouver. A collaborative round table of interested parties was established comprised of scientists, representatives from the provincial and federal governments, BC Hydro, First Nations and conservation interests. The water use plan for the reservoir system was drafted in 1996 and recommended a substantial reduction of discharges for power production from 200 cfs to between 90 and 105 cfs depending on storage levels to improve fish productivity in the rivers below the reservoir.

Given the negative publicity of the drawdown during the election campaign and the support that BC Hydro received for restoring some of the natural capital in the Stave-Allouette system, the two governments and BC Hydro initiated a formal program to complete water use plans that balanced power production with social and environmental values across all of the 23 watersheds that collectively contain the 34 generating facilities. The principles guiding these plans support many of the more generic principles that underlie sound governance for sustainable water resource management:

- Formal recognition of multiple objectives for economy, environment and social values
- No change in existing constitutional and legal rights
- Collaborative and transparent decision making processes
- Science-based decisions using adaptive management techniques
- Consultation and accommodation of First Nation interests and rights under the constitution

The program ran for five years (1998-2003) and cost BC Hydro $ 26 million to fund both the science (monitoring, fish habitat assessments and hydrology) and the round table consultative process. All but one of the water use plans resulted in a consensus across a broad range of interest groups including the First Nations. The provincial government established a Systems Operating Fund originally totaling $3.6 million to pay BC Hydro for forgone power benefits associated with re-balancing discharges to met environmental and social values. This fund has since been increased to approximately $40 million a year to pay for the plans across all 23 watersheds. The costs will ultimately be borne by rate payers as BC Hydro has been granted a rate increase from the independent BC Utilities Commission that regulates the utility.

There have already been substantive benefits to this multi-objective approach to reservoir operation. All litigation initiated by DFO has been dropped saving millions of dollars in legal fees and uncertainty over the operating parameters for managing reservoirs by the federal agency. Canada won the challenge under NAFTA on the basis that the water use plans led to the true cost of generating power and protecting environmental and social values. The constitutional responsibilities of the provincial government to consult and accommodate where necessary First nations asserted claims to rights and title have been met. Technically the water use plans are recommendations to the provincial government to amend BC Hydro's water license which sets up the legal entitlement for the utility to operate its reservoirs. BC Hydro's participation in water use planning is voluntary because it retains its legal rights to water. No formal compensation has to be negotiated when these rights are diminished to meet new environmental standards but arguably the increase in the rate structure represents a shift towards full cost accounting by the rate payers to regain some of the natural capital associated with the new operating practices for these reservoirs. Fisheries experts predict that productivity will substantially increase below the reservoirs, but the water use plans have a built in requirements for adaptive management so that additional measures can be undertaken if future monitoring deems this to be required.

The larger canvass for sustainable resource management

The water use planning initiative represents a relatively restricted form of collaborative decision making , involving principally the two levels of government, the First nations and a crown corporation. Attaining sustainable water resource management will require the engagement of all the four institutional domains that represent the governance framework outlined in the introduction to this paper and a much more diverse set of resource management issues. Such innovative approaches to governance and sustainability are occurring in the Fraser Basin and on the Central and North Coast of the province. This section of the paper outlines some of these innovative approaches.

The Fraser River Basin

The Fraser River is the fifth largest basin in Canada, covering 238,000 km2 roughly the size of California (See figure 1). It faces major challenges to long term sustainability, notably associated with the rapidly growing population of the Lower Fraser Valley around Vancouver; the loss of biodiversity in the wetlands in its estuary due to urban growth and agricultural development; water temperature increase approaching lethal limits that threatens the sustainability of pacific salmon, which is the mainstay of the province's commercial and recreational fishery; increasing probability of major flooding in the Lower Fraser Valley due to greater fluctuations in precipitation and snow accumulation plus increasing occurrences of drought and forest fires in the Central Interior of the basin due to general increases in summer temperatures over the past decade.

Governance of water and related resources is split across three levels of government and further diversified within each of these levels by a number of agencies (See Table 1). This multi- faceted approach to governance does not lend itself to sustainability as decisions are made by individual agencies in accordance with their specific mandates and do not reflect the integrated nature of the ecosystems which comprise river basins.

As an innovation in governance and in recognition of this compartmentalization, the two senior levels of government in the early 1990's created the Fraser Basin Management Board. This Board became a partnership between the three levels of government, First Nations and civic society leaders to develop a sustainable management plan for the basin. Over its five years in operation, the Board created a vision - "the Fraser Basin is a place where social well-being is supported by a vibrant economy and is sustained by a healthy environment." It also created a Charter for Sustainability that set out 12 principles for sustainable management of the basin (Fraser Basin Council, 1997).

In 1997, the two senior governments endorsed the Charter and vision statement and supported the establishment of a unique not- for -profit society called the Fraser Basin Council. The entity consists of a Society with representatives for the three levels of government, the First Nations and civic leaders who are responsible for the Society's constitution and bylaws and appoint the chair and vice- chair of the council's Board of Directors. The Council Board comprises of 36 members, is the custodian of the Charter for Sustainability and undertakes a wide range of collaborative initiatives that would be difficult for the traditional forums of government to undertake because of the "silo" approach to water resource management.

Two examples illustrate this approach. For a number of years gravel has been accumulating in the bed of the Fraser River where it discharges into the flat flood plain below the town of Hope, located about 100 km east of Vancouver. This bed aggregation has raised significant concerns with the local governments in the area about increased flood risk, public safety and erosion of river banks and dykes. Under formal agreements with the two senior levels of government which financed the flood control infrastructure three decades ago, local governments are wholly responsible for financing the maintenance these facilities. Gravel removal has been blocked by DFO because of its mandate to protect fish habitat. There are additional complications associated with some of the First nation bands in the area who claim aboriginal title to the sand and gravel. This impasse continued for five years though considerable scientific study was undertaken on river hydrology, gravel movements and there were some small trial removals of gravel to assess impacts on fish habitat.

The Council was the only body that could facilitate a solution to this issue as its directors contained representatives of the four levels of government involved in the dispute as well as community and environmental interests. On September 14, 2004 a five year agreement was signed by the ministers representing the two senior levels of government with approval to remove 500,000 cubic meters of gravel for each of the first two years and 420,000 cubic meters for the next three years. It is estimated that the annual discharge of gravel in to this part of the basin is between 300,000 and 350,000 cubic meters so this agreement will actually reduce the bed load over the next five years thereby lowering the flood risk. The agreement has many of the hallmarks of sustainability as it increases public safety, provides an economic base for surrounding communities through the sale of the sand and gravel, applying the provincial royalty for the gravel to flood management and maintaining environmental standards for fish management by developing a specialized hydrologic model to predict changes in river flows and velocities resulting from gravel removal based on best available scientific knowledge.

The second example is the Council's approach to developing sustainability indicators to track progress toward sustainability across the basin. Since 2002 the Council broadened its approach to sustainability by creating its first report on 16 indicators covering the economic, social and environmental health of the basin. The Council released its "Snapshot On Sustainability: State of the Basin Report" in January 2003. The 16 indicators can be sorted in accordance with the four pillars of sustainability mentioned in the Introduction:

- Environmental management: flood risk assessment; species at risk; climate change including greenhouse gas emissions; water quality and air quality
- Social Values: housing, education ,health
- Economy: economic diversification, employment, income, energy, agriculture, forestry
- Governance: corporate social responsibility, aboriginal and non aboriginal relations; community engagement – the level of participation by civic groups in decision making forums in the basin.

Over 500 experts participated in the development of these indicators and nearly 10,000 copies of the report have been distributed to government organizations, agencies and individuals.

Further evolution of these sustainability indicators was presented at the Fraser River Basin workshop held November 2004. Progress or lack of progress in the 16 indicators was tracked, and there was a more in depth analysis of indicators that affected by the Council's strategic initiatives such as improvement in aboriginal and non- aboriginal relations, economic diversification for communities, sustainable fisheries management and flood control including financing flood infrastructure. In addition one of the regional councils within the basin has prepared its own state of the sustainability report which will be a pilot for more in-depth assessment for this region and the other four regions in the basin. The success of this initiative is the degree to with the Council is able to influence the range of governance bodies in the basin to amend existing policies and programs to make further progress towards sustainability.

Central coast land and resource management plan

In the Introduction to this paper, governance was defined to encompass four institutional domains which now influence public policy-making, namely the traditional levels of government represented by the public sector; the corporate sector; the civic sector and the knowledge sector. The principal thesis of this paper is that sustainability is best advanced when these four domains collaborate in making public policy decisions.

The author was directly involved in the application of this thesis in the completion of a land and resource management plan (LRMP) for the Central Coast region of the province. This planning initiative does not emphasize water management per se but its principles and processes should provide useful information for water managers.

The planning unit covers 4.6 million hectares of the central coast of British Columbia stretching form the northern tip of Vancouver Island to Princess Royal Island south of Kitimat (see Figure 2). The main thrust of the plan was the protection and management of old growth temperate forest which drapes the coast and represents some of the last remaining large intact watersheds for this ecosystem globally. In 2000, international environmental groups influenced major wood retailers to boycott timber and pulp products originating from the old growth forests on the central coast until sustainable forest management practices were in place. The timber companies with cutting rights to the timber formed an unprecedented alliance with the environmental organizations- called the Joint Solutions Project (JSP) - to encourage the provincial government to set aside approximately one third of the old growth forest from forest harvesting until the LRMP was completed. The government agreed to an initial plan in April 2001 supporting this recommendation though this plan was not supported by many of the resource interests, local government or the First Nations. It asked the parties to complete the plan for the coast by the end of 2003 on a consensus basis if possible.

To facilitate a sustainable solution for the LRMP a coast completion planning table was formed with representatives from the corporate sector- forestry, mining, tourism and labour; the conservation sector; the First Nations bands – there are 20 bands representing the majority of the resident population in the planning area, and the local and provincial levels of government. The Table was informed by five sets of information:

- The 2001 initial planning agreement with a draft land use plan and information base
- New science- based information on eco- based system management (EBM) on forest and other resource uses provided by an independent science team many of whom came from the knowledge sector
- First Nation values, traditional uses and interests associated with asserted aboriginal rights and presented in some cases on land use maps
- Socio- economic assessments of various land use zoning scenarios to evaluate impacts on annual timber harvest, provincial revenues, income and employment.
- Presentations by the non- aboriginal sectors on the values and interests for land and resource use.

The Table met over a dozen times and incrementally fashioned the plan around the following principles for EBM.:

- Aboriginal rights and title are recognised and accommodated
- Ecological integrity is maintained
- Human wellbeing is promoted
- Culture, communities and economies are sustained within the context of healthy ecosystems
- The precautionary principle is applied
- Collaboration in reaching recommendations for government
- People have a fair share of benefits from the ecosystems in which they live.

The plan was completed on time and it met all the above principles. Forestry was supported on two thirds of the land base and mining on an additional 12 percent of the planning area. Forestry and mining were to follow the rules for EBM which required a higher level of environmental stewardship that on the other coastal forests in the Province. A similar agreement was reached for the North Coast LRMP which covers 1.7 million hectares contiguous to the central coast using the same collaborative model for reaching consensus. The environmental sector proposed an imaginative funding mechanism that would provide $60 million funding from conservation foundations for a new Coast Opportunities Fund provided that the two senior levels of government collectively matched this amount and the environmental standards associated with the land use zoning map portrayed on Figure 2 and EBM logging was implemented. The trust fund would be largely directed to funding aboriginal communities to become economically sustainable. The Planning Table recommended a no-net job loss commitment by government as a result of the plan with flexibility that if the EBM approach for logging and mining was overly restrictive some of the environmental management targets could be adjusted to maintain the commitment to this economic agreement. The Table also proposed an EBM Working Group made up of representatives from First Nations, the forest industry, the conservation groups and local government that would track the application of EBM measures to the forest resource over a five year transition period to ensure that the principles of EBM were achieved. This group would be advisory with the final decision resting with the Provincial government in consultation with the First nations.

The recommendations from the planning table are now being negotiated between the Provincial government and the First Nations as is required under the Canadian constitution. Similarly there are ongoing discussions on the establishment of the Coast Opportunities Fund. But there remain significant challenges to implementing this plan. The application of EBM principles shifts the focus of resource management to maintaining the functionality of forest ecosystems while accommodating commercial logging. This shift in paradigm will set up a clash between the application of precautionary science and the economic reality of commercial logging in mountainous terrain. First Nations will play a key role here as they desire economic opportunity for their communities and will challenge conservative application of the precautionary principle proposed by environmental groups. However agreements to date are a good example of convergence in governance among the four institutional domains noted earlier. Successful outcome for the plan when implemented will be sustainable forestry, restoration of degraded habitats, increases in employment due to market certainty for all resource sectors and an ongoing collaboration in governance that binds the key players to work through the transition to a more restorative economy on the central coast.

Future direction for governance and sustainability

This paper has outlined a few approaches to collaborative governance in which the author has directly participated. However these initiatives are well short of a well integrated approach to governance and sustainable water resource management in a province that faces the twin challenges of increasing demands for water and a potential reduction in supply due to climate diversity, increasing frequencies of droughts and rising temperatures in streams affecting both commercial and recreational fisheries. A much more integrated approach to sustainable management is required that links strategies for demand management, protection of natural capital, pricing policies and regulatory changes that enforce metering for conservation infrastructure in residential and industrial sectors. In this province less than half of the communities have metering and very few have block unit pricing to encourage conservation.

The five principles that have underlain the tests for sustainability in this paper should be used to check future approaches to sustainable water resource management:

1. **Living within Environmental Limits**—maintaining and restoring ecosystems to their proper functioning condition and ensuring that they remain so for future generations.
2. **Achieving a Sustainable and Growing Economy** – by respecting nature's boundaries and by continuously adding value to the use of natural capital.
3. **Ensuring a Stable and Just Society**—where there is social cohesion and inclusion in decision-making and by creating equal opportunity.
4. **Using Sound Science**—by applying independent science to inform decision makers but taking into account uncertainty and changing public values.
5. **Promoting Good Governance**—by actively promoting effective participatory systems of governance across the four institutional domains outlined in this paper.

Such comprehensive approaches have been initiated in other Provinces notably Alberta and Manitoba. The "Water for Life" program recently announced by the Province of Alberta addresses most of the points of eco- efficiency noted above encourages collaborative management in major watersheds through stewardship groups and links land use and settlement planning with watershed management. In a similar vein as the State of Victoria's imaginative policy "Our Future Our Water," Alberta has proposed transferable water rights for irrigation that can be sold in the market place to encourage more efficient use of water. Alberta does not contemplate the concept of "environmental water reserves" as proposed by Victoria State where water is retained in streams to preserve biodiversity values, fisheries, recreation and water quality. Arguably the water use planning initiative undertaken by BC Hydro provides such a reserve through amendments to the water entitlements held by the power utility.

As Paul Hawken noted in his book on "natural capitalism" (Hawken,1998) most of the technologies and governance models required for sustainability are already in place in various combinations throughout the world. The key is to energize the imagination of the four institutional domains that now form the governance framework for our resource values in a way that continually improves the level of sustainability before we lose the option of maintaining our natural capital by incrementally applying eco-efficiency tools faster that the pace of population growth. The hallmark of sustainability is to support and strengthen restorative behaviour to encourage long term solutions based on ecological principles rather than short term gains based on financial and economic motives. Ecological restoration can best be carried out by smaller, more flexible enterprises than by the larger bureaucratic organizations. The incentives set by governments should be steered to unleash the imagination, courage and innovation of individual citizens and the private sector. These levels of ecological thinking must be developed in all levels of society not just the political elite. This is the collaborative process that worked on the Central and North Coast planning tables and has supported the more progressive initiatives in Alberta and the State of Victoria.

Sustainable water resource management needs this type of paradigm shift to entrench the five principles into decision making in a way that still provides economic gains for communities and individuals but that fundamentally reduces the footprint of human use of resources. This paper has touched on only a few examples of innovative governance that is beginning to turn the tide. These examples need to multiply to become universal if water resources are to be sustainably managed. A number of governance functions need to change to make this happen. First government structures need to adapt to deal with ecological principles and look at whole watersheds not deal with the component parts through individual mandates that drive traditional departments. Second, governments need to introduce universal policies on pricing and regulations to encourage eco-efficiency for both the corporate sector and individual citizens. Third civic groups such as the Fraser Basin Council need to take a more proactive stand to encourage sustainable resource management through using the indicator reports to point our clearly where current policies are inadequate. And finally the education system needs to reorganize its curriculum to encourage the development of ecological citizenship in which today's youth are educated in the principles of eco efficiency and sustainability. Such educational programs should become mandatory for the next generation (Dobson, 2003) so that as consumers they can influence the market place to encourage trading only in sustainable commodities. When all levels of governance from the international organizations down to the local level collaborate with the other three domains of governance there is optimism that water, surely our most precious resource can on the longer term bc sustainably managed.

References

Dobson, A.(2003) *Citizenship and the Environment.* Oxford University Press.

Department of Environment and Sustainability, Victoria State Government. (2004) Our Future Our Water.

Frankel,C (1998) In Earth's Company. Business, Environment and the Challenge of Sustainability. New Society Publishers.

Government of Alberta, (2003) Water For Life.

Hawken, P., Lovins, A. and Lovins, H. (1998) *Natural Capitalism: Creating the Next Industrial Revolution.* Boston, Little Brown.

Pierre, J. (ed.) (2000) *Debating Governance. Authority, Steering and Democracy.* Oxford University Press.

Willard, B. (2002) The Sustainability Advantage: Seven Business Case Examples of a triple Bottom Line. Gabriola Island, British Columbia: New Society Publishers.

Privatisation of municipal waterworks and sustainability of water: a case study in eastern Jakarta, Indonesia

N.E. Shofiani* and J.E. Gustafsson**

*Jalan Kemuning IV No. 4 Blok C, Cihanjuang Indah-Cimahi, 40513 Jawa Barat, Indonesia (E-mail: nurendah@yahoo.com)
**Department of Land and Water Resource Engineering, KTH. Brinellvägen 28, SE-100 44 Stockholm, Sweden (E-mail: janerik@kth.se)

Abstract Privatisation is an important trend in global water arena. At least 93 countries have partly privatized their water service. Privatisation of the Jakarta Municipal Water Company (PAM Jaya) was started in 1997 when concession contracts were signed for 25 years with RWE Thames Water and Suez-Lyonnaise des Eaux. With a local partner, Thames Water formed a consortium PT Thames PAM Jaya (TPJ), providing service to the people in the eastern half of Jakarta. The other half is operated by PT PAM Lyonnaise Jaya, a consortium of Suez-Lyonnaise des Eaux. This paper evaluates the privatization in eastern Jakarta and specifically examines the current impacts and the future implications with particular emphasis on the consumers. The study shows that the major problem in the privatisation process is that Indonesia was rushed into it, without any existing regulation framework, resulting in a premature institutional transformation and invalid procedure. The two private companies were assigned by means of direct appointment due to the absence of a bidding process. Lack of public participation and consultation during the initial process of privatisation had eliminated the role of the public to decide the rights to water and the obligation of the government to provide the most fundamental basic needs of the people – water at an affordable price. After five years of TPJ's operation, there has been a slight change in water continuity in some area, but the problems of water quality and quantity remain.

Keywords Consumers; drinking water supply; Jakarta; PAM Jaya; privatisation; TPJ

Introduction

At the end of 1994, only 36% of Indonesia's urban population of 67 million had access to piped water. A huge amount of capital needs to be invested in the water sector to achieve the objective of serving 62% by the end of 2008 (JICA, 1997) considering that most of the Municipal Drinking Water Supply Enterprises (Perusahaan Daerah Air Minum – PDAM) were unable to provide such funds. To attract private investment, the Government of Indonesia (GoI) introduced the transformation of PDAMs into customer-oriented service industries. Furthermore, the GoI has initiated to set up water service fees based on the economic good principle rather than based on socio-political concept that treats water as a public good (World Bank, 1997).

DKI Jakarta (the special capital region of Jakarta) is the capital city of Indonesia where approximately 662 km^2 has a status of special territory in which the Governor is the head of administration. The population of Jakarta in 2000 was 8,385,639 with the growth rate of 0.16% during 1990–2000 (Statistics DKI Jakarta, 2002). However, the population rises up to 11 million during the day with people commuting from suburban areas for work.

In Jakarta, surface water and ground water resources are reaching a critical stage of their maximum use due to the high population. The water quality has declined owing to massive industrialisation, urbanisation and lack of government control. The need for water supply in Jakarta by 2005 is estimated at 31.5 m^3/s with an average per capita consumption of 150–200 litres per day. According to the Study of the Revise of Jakarta Water Supply Development Project (JICA, 1997), the target of Jakarta's water supply development is to increase the piped water supply for household and other socio-economic activities, to decrease the ground water abstraction and to have an even distribution and an affordable water supply system.

In 1995, PAM Jaya[1] was only able to serve 39% of the population, covering 426,735 house connections. Meanwhile, the rest of inhabitants still used groundwater as the source of drinking water. To improve the service, the GoI attempted to invite two private partners to participate in Jakarta's drinking water supply, RWE Thames Water and Suez-Lyonnaise des Eaux.

The following sections will discuss more on the public private partnership (PPP) in drinking water supply in the eastern half of Jakarta to evaluate the performance of the private partner, PT Thames PAM Jaya (TPJ), during the first 5 years of its operation and to describe the stakeholders' perspectives, most importantly the consumers.

Methods

This research is based on an inductive methodology, where observations were used to generalize findings. A triangulation approach was applied to validate the data by: (1) literature review, including reports, journals and other published documents as well as documents provided by the interviewees; (2) in-depth interviews during July–September 2003 with representatives of six stakeholders groups: government institutions, regional water authorities (PAM Jaya and the Regulatory Body), international financial institutions (IFIs), private company (TPJ), non-governmental organizations (NGOs) and associations; and (3) a consumers' survey in Eastern Jakarta during July–August 2003 (Shofiani, 2003).

Both the stakeholders' interview and consumers' survey were conducted by addressing standardized questionnaires adopted from Suleiman (2002) and modified according to the local situation. However, the in-depth interview as a method of qualitative research was based on pre-structured but necessarily unstandardized format of questionnaire with most questions being open-ended whereas the consumers' survey was based on a standardized questionnaire of definite questions.

The consumers' survey was a method of gathering information from several individuals who represent a larger population that can be analysed, patterns extracted and comparisons made (Bell, 1999). To allow statistical generalization, random samples are considered suitable (May, 1997). However, according to Bell (1999) samples are best stratified based on certain characteristics to ensure that the sample of population is truly representative. A significant characteristic used in this research was the proportion of respondents, which is similar to the actual proportion of consumers. Those characteristics are the number of consumers in each area (rayon), tariff groups, and the ratio of men and women observed.

[1] Jakarta Municipal Waterworks: the only authorized PDAM in Jakarta.

Public–private partnership in Jakarta drinking water supply

Privatisation can be partial, leading to so-called public private partnership (PPP); or complete, leading to total elimination of government responsibility or divestiture (Gleick *et al.*, 2002). In the case of Jakarta, PAM Jaya as a public water utility awards private companies the responsibility for operation and maintenance with a long contract model of 25 years (full concession), but asset ownership continues to reside in PAM Jaya. Thus the privatisation of PAM Jaya is classified as partial.

Pre-privatisation situation

Water sector activities within the national plan are treated on a sub-sector basis and fragmented among ministries and agencies that do not communicate with each other. According to the World Bank's report in 1997, this may lead to competing priorities and inconsistency among program proponents. The complexity of institutional arrangement in water supply has resulted in responsibilities that are unclear and confused (ADB, 2001).

During the period 1945–1968, the Public Work Agency was responsible for the Jakarta drinking water supply. PAM Jaya was officially established as an independent water company owned by PEMDA Jakarta (the provincial government of Jakarta) on 30 April 1977. PAM Jaya reported and consulted their development plan with PEMDA Jakarta. Decisions made by PEMDA Jakarta should be discussed with DPRD Jakarta (Jakarta City Council) before approval.

According to JICA (1997) PAM Jaya experienced serious organisational and technical problems. Overstaffing resulted in various organisational difficulties such as overlapping and irrelevant task assignment between organisational components. Lack of inter-departmental cooperation and coordination and excessive centralization of decisions resulted in delaying some services to the consumers. From human resources viewpoints, ineffectiveness was mainly caused by unbalanced personnel distribution in terms of age, place of assignment, specialisation, ineffective punishment and reward mechanisms, and unclear carrier path systems.

In terms of technical problems, total production capacity of PAM Jaya in 1997 was 498 million m^3 but only 47% was distributed to consumers. There was 53% of non-revenue water due to leakage from the distribution system, illegal use and billing errors. The numbers of connections increased rapidly by 169% for a decade up to 485,336 in 1998, where 231,607 are in eastern Jakarta and 253,729 are in the western area. However, this covered only 26% of 1.9 million households in Jakarta. According to JICA (1997), there are 5.8 persons per connection.

To accelerate the service improvement a new management system needed to be introduced to replace or rehabilitate old pipes and installations. PAM Jaya required a big budget, while its ability to provide the funds was limited. An alternative of involving private sector was considered the best solution to cope with financial difficulties. The private sector was expected to introduce professional management system and increase efficiency, which would benefit the consumer.

Privatisation history of PAM Jaya

Following the announcement of the President of the Republic of Indonesia on 5 June 1995 and the subsequent announcement by the Minister of Public Works, Jakarta water supply was divided into two concession areas, Jakarta East (zones II, III and VI) and Jakarta West (zones I, IV and V), with the River Ciliwung being the boundary (Figure 1).

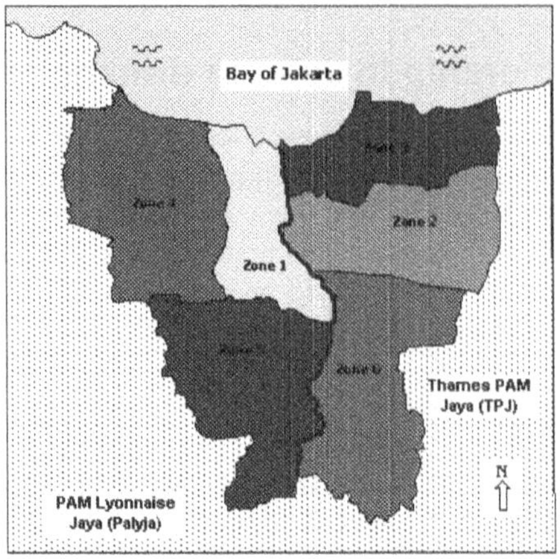

Figure 1 Jakarta water service area and the zoning system

Two private companies, RWE Thames Water and Suez-Lyonnaise des Eaux, were appointed to undertake comprehensive functions, which include operation, maintenance, management and the development of water supply system in the respective areas for 25 years. TPJ as a consortium of Thames Water and a local partner, PT Kekarpola Airindo is responsible for Eastern area of Jakarta and PT PAM Lyonnaise Jaya (Palyja), a consortium of Suez-Lyonnaise and PT Garuda Dipta Semesta is responsible for the Western area.

PAM Jaya and the respective consortia initially reached an agreement and signed the Memorandum of Understanding (MoU) on 6 October 1995. The two consortia were obliged to provide a feasibility study within 4 months. The final contracts were officially signed on 6 June 1997 and came into effect on 1 February 1998.

As political and economic crises coincided with the start of the agreement, the implementation of the partnership became restricted. Some stipulations could not be achieved, particularly the projected financial assumptions. Therefore on 1 June 1998, just 3 months after the kick-off date, all parties agreed to renegotiate the partnership agreements. The two most important components were reviewed: the water charge and technical targets. Finally the amended cooperation agreement was officially signed by the governor of DKI Jakarta on 22 October 2001.

Implementing private water supply

Privatisation of Jakarta water has created many controversies. The involvement of IFIs forced the GoI to adjust its policies to comply with global challenges. The GoI initiated to treat water according to the economic good principle, which means the allocation of water will primarily be based on its competitive uses to maximize the benefit from water and the concept of full cost recovery. This is likely to increase the water tariff which may cause limited access to water for the poor. Furthermore, the economic good principle may lead to a decline in the quality of life, since water is a major factor in supporting the public health.

There was no deficit on PAM Jaya's operation before privatisation (Surjadi, 2003). PAM Jaya, according to Indonesian Corruption Watch, had achieved the status of self-reliance and annually donated IDR[2] 3–13 billion to PEMDA Jakarta Revenues. A big investment loan was awarded by the World Bank to build two large-scale treatment plants and the expansion of distribution network during the period of 1992–1995. Soon after the construction was

[2] IDR: Indonesian Rupiah, as of June 2003 USD 1 = IDR 8,250

finished in 1998 and the networks were finally done, the privatisation process took place. Asset and operational management were all transferred to TPJ and Palyja. PAM Jaya had no chance to use the assets built with the loan and the private partners reaped all the benefit of new installations.

Regulatory framework and privatisation process

According to Beecher (1999) and Gleick *et al.* (2002), the advantage of privatisation may only be realized if coupled with an integrated regulation, a competitive bidding process, a performance review, a mechanism of conflict resolution and a skilful contract negotiator. In the case of Jakarta, it was all the opposite: (1) there was no regulatory framework as the basis for PPP when the idea of privatising Jakarta water was first initiated in 1995; (2) the two consortia were assigned by direct appointment governed by vested political interest; neither a public consultation nor a bidding process were performed in selecting the concessionaires; (3) the private partners were skilled at negotiating favourable terms of contracts as proven by the amendment of the original contract agreement in 2001 where many of the targets were set up lower than the original ones; and (4) there was no Regulatory Body to monitor the contract and mediate any dispute at the time of the initial cooperation agreement; the Regulatory Body was established in 2001.

The Regulatory Body's mission is to balance the interests of private and public sectors. Working properly it should play an important role in supervising the implementation of the contract, facilitatng conflict resolution between parties, adjusting tariff, coordinating other public institutions, and monitoring the performance of the private partners. However, the Regulatory Body has very limited power and mainly deals with technical issues.

The regulation on cooperation guidelines between PDAM and private sector[3] was issued in 1998, a year after privatising PAM Jaya. A higher regulation regarding PPP on infrastructure development and/or management[4] and its implementation guideline[5] was issued in 1998. A more specific guideline regarding PPP on the provision and management of drinking water and sanitation[6] was only issued in 2002.

Roles of international financial institutions

The role of two donor agencies, the World Bank and the Asian Development Bank (ADB), in the water resources and irrigation sector in Indonesia is very significant. In water supply, the World Bank's involvement started in 1974 by providing loans for water supply projects in five cities. The concept of cost recovery and efficiency pricing in the public sector was firstly introduced by the World Bank in the Second Water Supply Project in 1979.

The World Bank developed a report on Indonesia Urban Water Sector Policy Framework in 1997 at the request of Indonesian Development Planning Agency (BAPPENAS). The report identifies major policy changes aimed at improving financial performance as well as making the water supply sector more attractive for private sector participation (ADB, 1999). The report was acting as the background paper for the World Bank to carry out a policy reform in water sector through Water Sector Adjustment Loan (WATSAL) provided in 1999. The output of WATSAL is the Law on Water Resources Management[7] promulgated in February 2004. It creates many controversies such as to promote privatisation to be in line

---

[3] Instruction of the Minister of Home Affair number 21 year 1996

[4] Presidential Decree number 7 year 1998

[5] Decree of the Minister of National Development Planning number KEP.319/PET/10/1998

[6] Decree of the Minister of Settlement and Regional Infrastructure number 409/KPTS/2002

[7] Law number 7 year 2004

with the loan conditionality. One major issue raised by the stakeholders is allowing private sector authorisation over water resource for commercial purposes. As the consequence, water users e.g. farmers and households are made responsible for the overall cost of water provision including profits.

At the moment, ADB is working on the establishment of a regulatory framework for private and public water supply and wastewater enterprises. There are four ongoing projects which have small components of privatisation. However, according to ADB's representative, privatisation is not the best solution. The GoI has to be very careful with the contract agreement and the service type. There are some services that should stay in public hands because the private sector is not willing or able to solve the problems of unserved areas on its own (ADB, 2003).

Water charge, water tariff and tariff shortfall

In the case of Jakarta two different concept of water charge[8] and water tariff[9] are applied. The private sector investment is compensated through the water charge, which implies that the government fully guarantees of any losses for the private company. The water tariff is set up based on the principles of cross subsidies between the high and low-income people and progressive rate or block tariff system. The lowest tariff block is 10 m^3 by assuming the average needs of water per person per day is 60 litres. But for the case of a metropolitan city like Jakarta, the assumption is far below the average consumption which according to JICA (1997) would be 150–200 litres per person per day in 2005.

The water charge is increased from time to time. In the original contract agreement, the water charge was fixed at IDR 1,303 per m^3 water while the water tariff was IDR 970 per m^3 water. It was increased up to IDR 2,543 in 2002. Figure 2 illustrates the gap of water charge and water tariff.

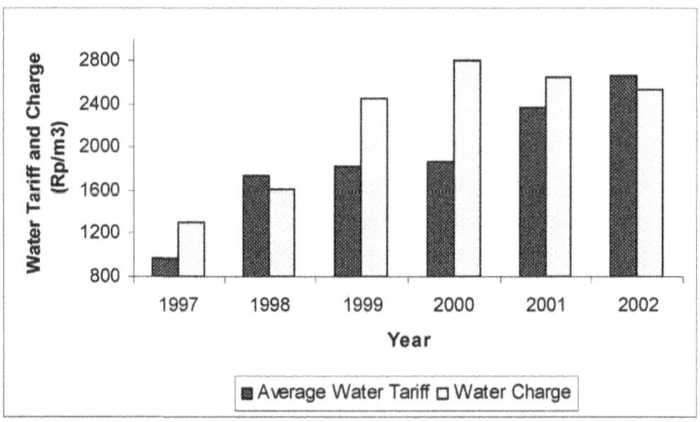

Figure 2 The gap of average water tariff and water charge

PAM Jaya had no other option but to raise the water tariff for the consumers to reduce the gap. Water tariff has been increased for over 250% compared with its pre-privatisation level and just in January 2004 it was quietly increased by 30%. Another argument for tariff

[8] The charge of water based on the value per volume (m^3) sold which is paid to TPJ to compensate its investment.

[9] The charge of water levied to consumers for customer service, including volumetric charge, fixed charge and meter charge.

increase was the current tariff level is still considered very low compared to the price of water sold to consumers by private vendors.

PAM Jaya is now increasingly indebted to the private companies since there is no alternative source of income after all its operation was taken over by the private company. In 2002, PAM Jaya had to cover the revenue shortfall to the private sector equalling to IDR 350 billion. PAM Jaya as the contracting organisation now believes that the privatisation has become just too expensive.

Target achievement

The achievement of the private sector could be examined by comparing the actual realization with the approved parameters: technical targets and service standards. In case of failure to meet the targets, the private company will be subjected to sanctions, penalties and even more, termination of contract. Among the five technical targets, only the volume of water billed is binding. Failure to achieve 70% of the targeted volume may lead to the termination of contract.

Along with the stipulation of the new agreement, the targets that are used today were set up at a lower performance level than the original one. Table 1 shows the comparison of the targets of the original cooperation agreement and the revised agreement during the first five years. The original targets were used during the period of 1998 to 2000 (Year 1 to Year 3).

Table 1 Technical Targets According to the TPJ Cooperation Agreements

Technical Standard	Volume Of Water Billed (million m^3)		Water Production (L/sec)		Unaccounted for Water (%)		Number of Connection (units)		Service Coverage Ratio (%)	
	1997*	2001**	1997	2001**	1997*	2001**	1997*	2001**	1997*	2001**
Year 1	105.00	91.96	8343	8523	50.00	58.07	251607	278083	55	57
Year 2	121.00	105.90	8531	7828	47.00	51.74	281607	285753	59	57
Year 3	136.00	117.94	8531	7408	42.00	45.86	311607	304303	64	59
Year 4	153.00	121.83	8531	7282	38.00	45.03	341607	315126	65	60
Year 5	186.00	131.32	8531	7309	35.00	43.03	361607	335413	70	62

Note: * Original Cooperation Agreement issued in 1997
** New Cooperation Agreement issued in October 2001

In 2002, TPJ has been able to meet 98–110% of the target in the new contract agreement. However comparing with the original targets, TPJ has only fulfilled the targets by 69–94%. Figure 2 describes the technical target achievements in 2002 according to the 1997 and 2001 standards in Eastern Jakarta.

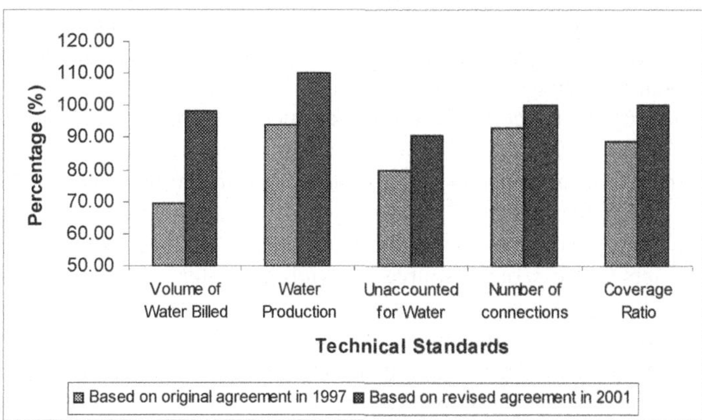

Figure 3 Target achievements in Eastern Jakarta in 2002

Perceptions and expectations of the consumers

Perceptions and expectations of the consumers were assessed by conducting a consumer survey. A total of 350 feedbacks from respondents was collected in the three zones of TPJ operating area.

General overview of the consumers

Total number of consumers up to end of July 2003 was 348,760. Consumers are classified into five different tariff groups based on their dwellings condition: social; very poor households; middle-low and upper-middle income households; and commercials. Low-income households are the major consumer group, which occupies of more than 65 percent of the population.

Based on the survey, the number of family members in each household varies, more than half of the households provide shelter for two families and have more than nine family members. Most of the consumers are working as employees in some private companies or are self-employed by creating an informal job of their own. Less than 3% have no permanent job and over 9% are working as private workers, such as labour or porter at traditional markets and ports; and drivers of *becak* (rickshaw) and motorcycle used for public transportation.

The monthly income distribution varies from less than IDR 500,000 to more than IDR 5 million. According to the Jakarta Governor Decree No. 302/2001, the minimum wage in Jakarta is IDR 425,250 per month. Based on the survey, more than 8% of the consumers have a salary of less than the minimum wage and nearly 65% of the consumers have a salary between IDR 500,000 and IDR 2 millions.

Major issues in drinking water provision

Consumers' awareness and knowledge on the water problems depends on their level of education. Basically this could be reflected by the monthly income earned. Better education level results in better income. Nearly half of the consumers are not aware of the privatisation of PAM Jaya. Most of them thought that the water supply system is still operated by a public authority or the government. It is obvious that there is a lack of public involvement and inadequate access of information as well, especially in the case of Jakarta where tariff cross subsidies are applied. Low income households that pay less than the high income group would obtain lower service level.

According to the survey, there is still unequal tariff distribution among consumers group. Nearly half of the low income consumers pay 10–20% of their monthly income for water. Most of the consumers refuse a further increase in water tariff since they have some difficulties in allocating their income for other basic needs as food, education, electricity and transportation.

There has been an improvement in water continuity as more than 60% of the consumers obtain water for more than 20 hours a day. However, a contradiction to this has also occurred. Water quantity decreases as the continuity increases. It is also recorded that nearly half of the consumers possess other source of drinking water, such as ground water with pumping system while the low income households must be satisfied by using shallow ground water wells since they cannot afford to install an electric pump.

In terms of water quality, consumers experience problem of water contamination due to leakages in the distribution pipe network. The water quality does not comply with the drinking water standard[10] as for several physical parameters: odour, turbidity and dissolved solids. Consumers have to boil the water before consumption. It means they have to pay more money in addition for buying water filters, paying for electricity or gas to boil the water, and they even buy bottled water as the tap water is unsafe because of bacterial contamination. It is recorded that one third of the consumers connected to the network consume bottled water.

From the consumers' standpoint, there has been no improvement in drinking water provision. The survey confirms that most consumers are not satisfied with the service. Their foremost opposition on water service is the high price of water, incorrect meter reading, delay in receiving water billed, which is subjected to penalties, poor water quality and slow response on consumer's complaint.

Environmental and sustainability issues of water privatisation

The privatisation of Jakarta water supply aims at increasing efficiency, which is believed to be the most economically and environmentally liable way to achieve water sustainability and to ensure safe water for the future. But, many critics argue that water privatisation is an unsustainable way of providing the most basic need for the people, because a private company will never have the social incentives to bring water for urban poor communities.

Profit orientation of the private company has put the price at an unaffordable level, resulting in limited access to potable water, which in turn will increase the use of groundwater. In 1995, registered private deep wells produced water at a rate of 0.96 m^3 per second. Nearly 88% of the total groundwater abstraction was unregistered and came from domestic use with a production rate of 7.06 m^3 per second (JICA, 1997). Land subsidence occurs due to excessive groundwater abstraction which is beyond its permissible level. As the result, the quality of groundwater has deteriorated by sea water intrusion, especially in northern part of Jakarta that lies in the coastal Jakarta Bay area.

The raw water for Jakarta mostly comes from rivers that are managed through three dams: Jatiluhur, Saguling and Cirata Dam. The water is also used for other competitive purposes such as power generation, irrigation, tourism and industry (Tutuko, 2001). In recent years the dry season flow has been reported to minimum levels (Tempo Interaktif, 2003: Pikiran Rakyat, 2004). The involvement of private profit-making companies might undermine comprehensive conservation efforts and fail to deliver on minimum water flow in rivers and streams that are needed to keep healthy aquatic ecosystem.

In addition to the environmental impact of water privatisation is the emerging business of bottled water utilizing springs. Consumers chose to buy bottled water for drinking and cooking purposes due to fear of bacterial and chemical contamination in the tap water. From ecological standpoint, spring water pumping is jeopardizing underground aquifers as the source of water for nearby streams, wells and farms. Water flow in the up stream is decreasing and water level is declining that would endanger the local biodiversity and further down stream ecosystem. On top of it, bottled water industries are accompanied by water bottle manufactures. Plastic bottles consume million of barrels of oil and most used bottles end up in landfill, adding to landfill crisis. In Jakarta, 2,840 m^3 of plastic wastes are generated every day or 11% of the total waste production of 25,600 m^3 per day (Trisyanti, 2004).

[10] Decree of the Minister of Health number 416/MENKES/Per/IX/1990.

Jakarta water privatisation, the next steps: a recommendation

Five years have passed since two concession contracts were signed by PAM Jaya, TPJ and Palyja. The overall achievements of the private partners have been far from satisfactiory; PAM Jaya continues cooperating with the private entities. The eventual termination of the contract or a legal dispute risks to be very costly. Moreover, PAM Jaya is obliged to pay the tariff shortfall to the private companies. It is worrying that when the concession term is over, it will no longer have the assets since these will be used to pay the accumulated debts (Nababan, 2004).

It is critical for PAM Jaya and the Regional Government of Jakarta to undertake important measures to counter the economic downfall as well as environmental damages. One possible solution is by empowering the Regulatory Body, which now *is* considered as a non-independent body. The Regulatory Body should be an independent body to act without any intervention from any dominating party and away from any political influence. The body should act only to protect the interest of the consumers and community, ensuring an equal and safe access to water for the entire inhabitants, regardless of their social and economic status. Both, urban poor as non profitable consumer group and high income people have the same right to water as a fundamental basic need.

The Regulatory Body is also responsible for estimation of water tariff and water charge. In this case, the regulator should be strict in evaluating any adjustment in water tariff and water charge as proposed by the private companies. More efforts should be made to consider the capability of consumers to pay for water and the ability of PAM Jaya to compensate the gap of water charge and water tariff.

The combined forces of PAM Jaya and the Regulatory Body might ensure the private partners pursuing a proper implementation of the contract. Technical targets and service standard should be strictly realized. Punishment and award mechanisms should be considered as an effective mean for the private partners to improve their performance.

However, the Regulatory Body would not be able to properly function without an integrated regulatory framework. Nevertheless, the establishment of a regulatory framework should be followed by a strong institutional capacity to ensure law enforcement and a proper court system.

If privatisation in Jakarta is to be continued as a sustainable option, it is important to increase the role of different stakeholders. Failure in implementing public participation, can cause ineffective service, unprivileged group discrimination and clean water violations. More efforts should be made on providing guidance on public access of information and monitoring and public participation and transparency.

References

ADB – Asian Development Bank (1999) Impact evaluation study of bank assistance in the water supply and sanitation sector in Indonesia.

ADB (2001) Regulatory systems and networking: water utilities and regulatory body. In *Proceeding of the Regional Forum,* Manila, 26-28 March.

ADB (2003) Beyond boundaries: extending services to the urban poor. Manila.

Beecher, J.A. (1997) Water utility privatization and regulation: lesson from the global experiment. *Water International* **22**(1), 13-20.

Bell, J. (1999) Doing your research project: a guide for first time researchers in education and social science (3rd edition). Open University Press, Buckingham, UK.

Brubaker, E. (2001) The promise of privatisation. Energy Probe Research Foundation, Toronto, Canada. [WWW document] URL http://www.environmentprobe.org/eniroprobe/pub/Ev548.htm

Gleick, P.H., Wolff, G., Chalecki, E.L. and Reyes, R. (2002). The new economy of water: the risk and benefits of globalization and privatization of fresh water. Pacific Institute for Studies in Development, Environment and Security, California, USA.
JICA – Japan International Cooperation Agency (1997) The study of the revise of Jakarta water supply development project (Final Report, Volume 2). Jakarta, Indonesia.
Lanti, A. (1996) Case study infrastructure privatisation in Indonesia: water supply DKI Jakarta. Research Triangle Institute. [WWW document] URL http://www1.worldbank.org/wbiep/decentralisation/eaplib/Lanti.PDF
May, T. (1997) Social research: issues, methods and process (2nd edition). Open University Press, Buckingham, UK.
Nababan, H. (2004) Water privatization: learning from Indonesian cases. Freedom for Debt Coalition. 24 March. [WWW document] URL http://www.freedomfromdebtcoalition.org/main/pages/000198.php
National Information Agency (2002) Indonesia 2002: An Official Handbook.
Pikiran Rakyat (2004) Water flow in Saguling is decreasing (Debit air saguling menyusut). 9 July. [WWW document] URL http://www.pikiran-rakyat.com/cetak/0704/09/0302.htm
Shofiani N. E. (2003) Reconstruction of Indonesia's drinking water utilities. Assessment and stakeholders perspective of private sectors participation in the Capital Province of Jakarta, TRITA-LWR M.Sc.thesis No 30, Department of Land and Water Resources Engineering, Royal Institute of Technology, Stockholm, Sweden.
Statistic DKI Jakarta (2002) Jakarta in figures 2001 (Jakarta dalam angka). Catalog BPS 140331.
Suleiman, R. (2002) Privatization of Jordan's capital water utilities: assessment and evaluation of water supply and wastewater service of Amman governorate. *TRITA-LWR M.Sc.thesis No 31*, Department of Land and Water Resources Engineering, Royal Institute of Technology, Stockholm, Sweden.
Surjadi, C. (2003) Public private partnership and the poor – drinking water concessions: a study for better understanding public private partnership and water provision in low income settlements. Water, Engineering and Development Center (WEDC). Leicestershire, UK.
Tempo Interaktif (2003) Dry season: water level in Jatiluhur Dam is decreasing 10 centimeters per day (Akibat kemarau, waduk Jatiluhur setiap hari susut 10 centimeter). 28 June. [WWW document] URL http://www.tempointeraktif.com/hg/nusa/jawamadura/2003/06/28/brk,20030628-01,id.html
Trisyanti, D. (2004). Solid waste management of Jakarta – Indonesia: An Environmental System Perspective. TRITA-KET-IM M.Sc. thesis No. 6, Department of Chemical Engineering and Technology, Royal Institute of Technology, Stockholm, Sweden.
Tutuko, K. (2001) Jakarta water supply. In *Proceeding Sustainable Urban Water Services – Hong Kong Seminar*, Hong Kong, 23-24 November.
World Bank (1997) Indonesia water supply sector policy framework.

The Water Reclamation Matrix: a framework for sustainable urban water use

Eric Rosenblum* and John Anderson**

*City of San Jose Environmental Services Department, 3099 North First Street, San Jose, California, 95134, USA. (E-mail: eric.rosenblum@sanjoseca.gov)
**New South Wales Department of Commerce, 2-24 Rawson Place, Sydney, NSW 2000, Australia. (E-mail: john.anderson@commerce.nsw.gov.au)

Abstract Drawing on the history of water reuse in San Jose, California (U.S.A.) a framework is presented that could help identify issues that must be resolved before a community is prepared to use recycled water. The framework consists of a matrix of key categories that affect the decision to recycle water from the perspective of each of the institutions responsible for producing and distributing recycled water. Environmental, technical, economic and political factors are identified and assessed as either facilitating or impeding a participating agency's implementation of a water reuse project. When numerical values are provided to indicate the relative importance of key factors as motivations for reuse, the matrix can also provide an estimate of the relative urgency with which a project might be adopted. This approach is similar in many respects to spreadsheet-type decision support tools used in Australia and elsewhere to compare and summarize the attitudes of the broader community of stakeholders towards alternative water reuse projects.

Keywords decision support; sustainability; water reuse

Introduction

Recycled water—wastewater treated to the degree appropriate for potable and nonpotable purposes—provides hundreds of communities around the world with a safe, reliable and environmentally sustainable water supply. Nevertheless, recycled water project managers in many cities in the United States continue to encounter resistance to water reuse. (Hartley, 2003) As indicated in Table 1, during the past five years several potable reuse projects were abandoned or deferred due to public opposition, and even nonpotable water reuse projects have spurred public concerns. Among the reasons cited by decision makers for rejecting reuse are health issues, environmental justice, a preference for other "cleaner" water sources, the inadequacy of existing regulations and the high cost of reuse.

Implementation of reuse projects is further complicated by the fact that more often than not they require the cooperation of a number of agencies, each of which is responsible for one segment of the urban water supply chain. As illustrated in Figure 1, the urban hydrologic cycle begins with a water wholesaler who collects water from natural sources and transports it to a water retailer, who in turn supplies it to the customers. Once used, wastewater is collected and conveyed to a wastewater treatment plant where a separate entity is often responsible for treating the wastewater prior to discharge or reuse.

Table 1. Recent U.S. water recycling projects modified or deferred

Location (Date)	Project	Stated Reasons for Opposition	Result	References
San Diego CA (1998)	Potable Reuse (MF/RO with reservoir augmentation)	Health concerns; environmental justice; cost	Deferred; renewed interest in 2004	Wegner-Gwidt, 1998; VanWagoner, 1999.
Upper San Gabriel Valley, CA (1996)	Potable Reuse (Tertiary treatment with groundwater recharge)	Health concerns; environmental justice;	Modified to limit recharge in vicinity of reservoir	Wegner-Gwidt, 1998;
Tampa, FL (1999)	Potable Reuse (MF/RO with reservoir augmentation)	Health concerns; environmental justice; cost	Rejected in favor of desalination of brackish water	CH_2M-Hill, 1993; Pinellas County, 2002
Dublin CA (2000)	Potable Reuse (MF/RO with groundwater recharge)	Health concerns; cost	Delayed; nonpotable allowed	Pleasanton, 1999.
Redwood City, CA (2003)	Nonpotable Irrigation (Tertiary treatment)	Health concerns; disagreement about shortage	Restricted from school grounds	DWR, 2003

Given the number of parties involved, even when the need for recycled water is apparent program managers are called upon to address a variety of concerns from multiple partners. This task is made somewhat more feasible if the manager understands the nature of these issues and their relative importance both to the participants and the project. An analytical method is described for evaluating the factors that influence adoption of water reuse projects in order to help managers respond appropriately and effectively to the needs of multiple stakeholders.

A matrix for evaluating factors influencing reuse

As illustrated in Figure 2, there is a hierarchy of factors that must be satisfied in order to implement a water reuse project. At bottom are the *environmental* drivers such as diminishing ground water supplies or degradation of downstream ecosystems that might be remedied by implementation of a water reuse project, as well as features of the physical setting that must be protected from project impacts. These drivers are satisfied when demands on the environment are managed in a sustainable way.

Technical issues are the traditional domain of engineers who prepare water supply plans and design project specifications. They include the ability of planned facilities to purify wastewater and similar problems that can be solved by the application of technology. Success in this category is a well-designed project, delivered on time and within budget, that meets water supply and wastewater treatment needs.

Economic factors include project benefits and costs and may even extend to the deeper question of what the project is worth to the community compared to other projects. On the plus side are the value of recycled water, the avoided cost of building and operating alternative water supply facilities and a variety of other market and non-market benefits. (Rosenblum 2003) On the minus side are the project costs including the lost opportunity to devote the same funds to other projects. Economic analysis also includes the question of how the participating agencies will pay their fair share and how they will distribute revenues from water sales. For the financial managers who work in this arena success is knowing that a project is a good investment that will provide returns to the community for many years.

Political considerations take into account all these factors as well as other issues that combine to influence decision makers and the public to support or oppose water reuse. For instance, the public may endorse water recycling because they are afraid of future water shortages or because they want to restore their local riparian environment. Or they may oppose a project in order to keep water rates low or to limit community growth. In each case the political consideration differs from the underlying environmental, technical or economic factor in that it also reflects the importance the decision makers place on the problem and the value they attribute to the solution.

Planners may be able to calculate the probability of water shortage with a confidence ±5%, but the importance of water supply as a political issue is not solely determined by such data. Instead, it reflects such intangible factors as the community's aesthetic appreciation of irrigated landscaping, their concerns about retaining water-intensive industry jobs, and which other critical need (e.g. housing, mass transit) currently occupies the headlines. Astute political leaders routinely consult with interest groups and poll their constituencies to find out how the community rates the seriousness of their problems and which solutions they prefer. As a result, for projects to succeed politically they must not only be well-designed and cost-effective, they must also satisfy the needs of the elected officials who approve them and the taxpayers who pay the bill. While these factors are often the most difficult to analyze, they usually determine whether or not a particular project will be adopted.

The matrix provides a framework for assessing a community's potential to implement a water recycling project by examining each category of issues facing the institutions responsible for each sector of the urban water cycle. Within each category, issues of concern to the participants are presented as inclining the agency either to support a water reuse project as indicated by a "plus" sign (+) or oppose it, as indicated by a "minus" sign (-). A neutral position within a category is represented by a "zero" (o). A completed column of "plus" signs indicates that an institution is aligned in favor of reuse, and may sponsor a reclamation project. When all participants are satisfied that the issues in each category are resolved to their satisfaction (a "plus" matrix) there is a high likelihood that all significant barriers have been removed and the project will move forward.

Managing a top-down process

Since values ultimately drive the decision process, the direction of influence in the matrix of decision factors is primarily form the top down. A project that has political support can be implemented even if other factors are not all favourable: with sufficient political will, economic resources can be found to purchase technical solutions to produce and distribute recycled water for a sustainable environment. On the other hand, without political support it is unlikely that a project will be implemented even if it addresses an environmental need in a technically and economically feasible manner.

Yet influence can also occur from the "bottom up," directly or indirectly. A critical environmental condition like an extended drought can quickly give policymakers sufficient motivation to overcome economic and technical challenges to water reuse. Less dramatically, a technological breakthrough (e.g. improvements in membrane design) can reduce project costs, simultaneously removing an economic barrier and appealing to a political leader's desire to be perceived as progressive and technically aware.

On the other side of the equation, an elected official may also oppose water reuse to curry favour with an electorate that is unconvinced about the safety of recycled water, politicising what is fundamentally a technical issue. Similarly, policy makers may oppose water reuse on the grounds that it is expensive without considering the full range of costs and benefits, turning an economic issue—lower water rates—into a campaign issue.

In all instances the upward influence between categories is essentially a process of translation whereby the consequences of an environmental, technical or economic event are expressed in terms of public values. When motivating factors can be clearly related to outcomes that effect the community's quality of life, then political leaders are much more likely to take progressive action. This was the situation in San Jose, California in 1990 when city officials voted in favour of implementing a regional water recycling project to prevent a limit on treatment plant discharge from halting regional growth.

Case study: South Bay water recycling (San Jose, California)

For nearly 20 years, from 1970 to 1990, the political leaders of San Jose, California (USA) considered and rejected a number of water reuse projects. Ultimately, in 1993 they made the decision to implement South Bay Water Recycling, a 60,000 m^3/day (15 mgd) regional nonpotable reuse facility. A brief review of the history of water recycling in Silicon Valley may help to trace the interactions of the various factors that influenced these decisions, and illustrate the use of the matrix as an analytical tool.

As early as the late 1960s the local water district considered extending a 30 km (20 mi) pipeline to a large federal water reservoir to increase local supplies. The leaders of Santa Clara Valley (as it was known before orchards gave way to computer companies) evaluated a number of alternatives, including agricultural water reuse, water conservation and additional local storage. The unit cost of the reclamation and pipeline projects were about the same (less than \$0.05/m^3 or \$60/af) but water reuse was rejected "out of concerns for public health," and the pipeline project was constructed at a final cost of about \$300 million or \$0.05/m^3. Different reuse projects were proposed over the next two decades, especially after the wastewater plant serving this area added multimedia filtration facilities in 1979, improving the quality of treated effluent to the point that it was suitable for nearly all types of nonpotable use. In each instance recycled water was rejected on the grounds that it was technically infeasible and more costly than available alternatives

However, a matrix analysis of the factors influencing decisions at that time (Figure 3) suggests that whatever technical difficulty these reuse projects may have posed, neither the water wholesaler nor the wastewater agency had any political motivation to recycle wastewater. Inexpensive, highly subsidized water was readily available from state and federal projects, and there was no apparent limit on the discharge of treated effluent to San Francisco Bay. From this perspective it is worth noting that as the price of potable water rose, estimates of the cost of recycled water also escalated, from \$0.05/m3 in 1976 to \$0.25/m3 (\$300/af) in 1981. Water retailers, for their part, appeared to have had little interest in these developments.

It is tempting to speculate that if the water and wastewater agency had been able to appreciate the combined benefits of reuse they might have supported a project. By working together, they were able to extract more economic value than was available to either agency alone. This also points to the larger problem of externalities—where benefits and costs related to a project occur outside of the agency's focus. (Sheikh 1998)

Both the political importance and the economic value of recycled water in Silicon Valley changed dramatically in 1989 when regulators proposed to limit wastewater treatment plant flows. The US Environmental Protection Agency and the California Water Resources Control Board observed that fresh water discharged from the treatment plant had over time converted salt marsh adjacent to the outfall channel to fresh or brackish marsh. Since the salt marsh was home to two endangered species, the Board threatened to limit palnt discharges to 450,000 m^3/day (120 mgd), less than the plant was receiving at that time. In response, the wastewater agency proposed to divert its highly-treated effluent by recycling up to 60,000 m^3/day (15 mgd), bringing to a close thirty years of deliberation on the merits of reuse.

In this case, the environmental issue was translated into a political issue by appealing to the public preference (embodied in the federal Endangered Species Act) that no flora or fauna unique to the United States should become extinct. In addition, the Board order also translated an environmental factor into an economic factor by increasing the value of water reuse which became equivalent to the least cost alternative project, a 20-km (13 mi) pipeline into the middle of San Francisco Bay estimated to cost as much as $150 million. (Rosenblum 1999)

At the same time that the wastewater agency was beginning to appreciate water reuse as an alternative to flow limits, the water district experienced a similar conversion due to an extended drought that had limited water supplies, inflating the cost of water and necessitating the imposition of unpopular restrictions on use. When water wholesalers also found the use of recycled water economically advantageous and politically attractive, the two agencies agreed to support a water reuse project.

At this point the water retailers, who had been largely silent, voiced their concern that the sale of recycled water would reduce their revenues. Although they did not oppose the project, they noted that state law granted them exclusive rights to sell water within their service areas. They also questionsed how the agencies would build and maintain a dual distribution system without interfering with their facilities. This situation is reflected in in Figure 4, below.

These issues were addressed during two years of meetings with water and wastewater agency staff, which established policies and procedures pertaining to the sale of recycled water in the retailers' service areas. These polices specified that 1) retailers would sell recycled water to customers and be kept "whole" financially by charging their customary markup, and 2) the responsibility for constructing, operating and maintaining the new facilities would remain with the public agencies. Once all parties agreed and the terms were documented in legal agreements, the project moved forward as reflected by the matrix presented in Figure 2.

Applicability and future work

As illustrated in the case study, the matrix can be used to categorize and identify issues that present obstacles to reuse. The matrix may also be extended to other types of environmental programs by selecting different participants according to their jurisdictions with appropriate factors influencing their decision (e.g. storm water management).

While the matrix presented in Figure 2 was explicitly developed to include only those entities with jurisdiction over a segment of the urban water cycle to assess their inclination to use recycled water, the matrix can be extended horizontally to cover additional stakeholders including water users, landholders, regulatory and financing agencies, community and environmental groups all of whom may have an interest in the project. In the same manner the matrix can be extended vertically to cover subsets of stakeholder issues within each of the four categories (i.e. environmental, technical, economic, and political).

Experience at the Department of Commerce in Australia also suggests that a matrix extended in this manner can be applied to help analyse choices where several project options are considered by multiple stakeholders considering a number of issues. To help participants rank projects according to their conformance with stated goals, the table in Figure 2 can be modified by substituting projects for participants in the top row and ranking project conformance with selected critieria which is substituted for the categorical factors along the left side of the table. Modified in this way, the matrix more nearly resembles a conventional decision support table that each participant can complete to rank projects. Such decision support tables have been used in stakeholder workshops which arrived at consensus rankings of the project issues, and are available as software packages programmed to summarize and average participant's views. (Rosenblum, 2000)

It has been observed that community stakeholders are remarkably adept at weighting the issues and working out the tradeoffs between project options, and they are equally skilled at combining project elements to develop new options that better meet community needs. In fact, given their commitment and creativity, it has often proved to be more important simply to have some kind of community participation process than to use some particular evaluation method. As the saying goes, "consult early, consult often"; it pays to identify stakeholders and their issues as early as possible in the process so that they can attain the same level of understanding about tradeoffs and choices and communicate knowledgably with project sponsors and with each other before the evaluation process is completed.

Calculating consensus

The process of consulting with a variety of stakeholder in a group setting is effective in part because it forces various "advocacy coalitions" to explain the basis of their concerns as a prerequisite for group process. Even agency stakeholders are capable of productive dialog and collaboration among themselves, provided that they are able to overcome the barriers inherent in their divergent perspectives. This can be achieved when individual members of stakeholder agencies communicate their "core" beliefs that underlie their "policy" beliefs (Sabatier 1988) and develop an understanding of their respective points of view.

The matrix above can be adapted to better reflect the importance of understanding between agency stakeholders by assigning scalar values (0 to 1 instead of +/- indicators) that indicate the significance to the agencies of the various motivating factors. With respect to water reuse, a value of 1.0 would indicate that an agency regards implementation of reuse as critical, based on the particular motivating factor, while a value of 0 would indicate that, with respect to that factor, the agency has no motivation to recycle water. Table 2 shows how the importance of categorical factors to the wholesaler, retailer and recycled water producer in the case study might be represented by their respective criticality index (K_1, K_2 and Kn).

As presented in Table 2, each participant's criticality index may also be estimated by the value of the political factor alone, since it is in some sense a cumulative factor and the greatest predictor of whether or not a project will be implemented. In the general case, the aggregate criticality is therefore represented as $K = \Sigma(K_1 + K_2 + ... + K_n)$ where K_1 through K_n are the political motivation of the participants. In this instance, the aggregate criticality is the sum of the political motivation values, or 1.6.

Table 2 Numerical Ranking of Participant Motivations (K=Criticality)

	K_1	K_2	K_n
$k_{Political}$	0.5	0.1	1
$k_{Economic}$	0.5	0.1	1
$k_{Technical}$	0.1	0.1	0.5
$k_{Environmental}$	0.1	0.1	1
K_P (Total)	1.2	0.4	3.5
$K_1, K_2 \ldots K_n$ (Individual Average)	0.3	0.1	0.9
$K_1, K_2 \ldots K_n$ (Political Criticality)	0.5	0.1	1
K (Aggregate Political Criticality)	1.6		

Criticality provides an important indication of the community's motivation to recycle water, but experience suggests that unless the participating agencies understand each other's motivations projects often get tied up in internal wrangling. An index of participant consensus can be calculated by comparing participants regard for each others' criticality measures. (Webb 2004) Table 3 presents the participant ranking of each other's political motivations as described in Table 2. In this instance all participants agreed with the wastewater agency's need to reuse water, but the wastewater agency had a higher regard for the importance of reuse to the other participants than either did themselves. Also, the water wholesaler and retailer each had a lower estimate of the other's need to recycle water.

Table 3 Average Ranking by Other Participants of Motivating Influences (Relative Criticality)

Self-Assessment	P_1	K_{P1}-K	P_2	K_{P2}-K	P_n	K_{Pn}-K	Sum	Average	K_r
$K_1 = 0.5$	**0.5**	na	0.1	-0.4	1.0	+0.5	0.9	0.45	K_{r_1}
$K_2 = 0.1$	0	-0.1	**0.1**	na	0.5	+0.4	0.5	0.25	K_{r_2}
$K_n = 1.0$	1.0	0	1.0	0	**1.0**	na	0.0	0.00	K_{r_n}
								0.70	K_r
								0.23	K_r/n
								0.77	$1-K_r/n$

As shown in Table 3, *relative criticality* (Kr_1 etc.) is the average difference between the self-assessment and the assessment of others, and ranges from 0 (complete consensus) to 1.0 (complete disagreement). In general, relative criticality $K_r = \Sigma(Kr_1 + Kr_2 + \ldots + Kr_n)$. Since Kr_1 etc. are between 0 and 1.0, their average is always less than 1.0; in this instance, the aggregate relative criticality is 0.70, and the average relative criticality (Kr/n) is 0.23. Subtracting this value from 1 yields the *consensus index* ($c = 1 - Kr/n$) which equals unity when all participants agree (i.e. $Kr/n = 0$) and approaches 0 the more they differ ($Kr/n = 1$).

Criticality (K) and relative criticality (K_r) can be used to estimate the time required to reach a decision. If the time (t) increases with the number of participants (n) and decreases with their motivation (m), this relationship is expressed as $t=f(n/m)$ (Equation 1). Motivation can be described as the product of criticality (K) and the consensus index (c) or $m = cK$ (Equation 2). Substituting 1- Kr/n for c in Equation 2 and cK for m in Equation 1 yields:

$$t = n^2/[K(n - K_r)] \quad (4)$$

Certain features of this equation are empirically appealing. For instance, since each participant must communicate with all of the others, it seems reasonable that the time to reach a decision should increase with the square of the number of participants (n^2). Similarly, it seems appropriate that the group's collective motivation (K) should be amplified by a factor (n-Kr) ranging from n (complete consensus) to 0 (complete disagreement) as Kr ranges from 0 to n. At the boundary condition where perfect consensus is achieved (Kr = 0) and each participant is urgently motivated to recycle water, then $K=n$ and $t =(n/n) =1.0$ representing the smallest unit of time required to reach a decision. When participants are all unaligned ($Kr=n$) then $K(n\text{-}Kr) = 0$ and they never agree to reuse water.

Applying this matrix-based formula to the case study presented, the authors suggest that the participating agencies collective need to recycled water can be gauged by the aggregate criticality K = 1.6, while the divergence of their opinions about each others' respective need to reuse is shown in the relative criticality index Kr = 0.7. Using the formula, the relative time required for the agencies to reach consensus is calculated as t = 2.4, suggesting that it took more than twice as long for those participants to reach a decision to reuse which conforms with the authors' observation.

This analysis is admittedly simplistic and incomplete compared to the more sophisticated analyses available in the literature of classical decision theory, many of which provide elegant models of multiple objective decisionmaking (Keeny and Raiffa 1976, Bell 1977). There are also a number of contemporary journals that address both the process and politics of decisionmaking in a more thoroughgoing manner (e.g. The Journal of Conflict Resolution from the Yale University Department of Political Science). Nevertheless, with respect to the goal of understanding a few factors that directly influence agencies to reuse water, this "mini-model" may be useful insofar as it suggests that participants can reduce the time it takes them to agree by ensuring that each has an accurate view of the others' circumstances. Other strategies may also be suggested by the equations which, notwithstanding their numerical values, are qualitative in nature. They are presented here only to highlight the utility of the matrix in clarifying the relationships between agencies attempting to implement sustainable water use.

References

Bell, David ed.(1977) Conflicting Objectives in Decisions. (New York: Wiley)

California Department of Water Resources (DWR). Recycled Water Task Force. "Meeting Minutes 2/26/03" http://www.owue.water.ca.gov/recycle/docs/Feb2603Minutes_Draft.pdf (Accessed 3/8/04)

CH_2M Hill. "Tampa Water Resource Recovery Project Pilot Studies." Tampa, FL 1993. Cited in *Issues in Potable Reuse: The Viability of Augmenting Drinking Water Supplies with Reclaimed Water* (National Research Council, Washington DC, National Academies Press) 1998

Hartley, Troy. Water Reuse: Understanding Public Perception and Participation. WERF Research Report 00-PUM-1. (Water Environment Reuse Foundation, Alexandria, VA) 2003 p.1-3

Kasower, Steve. Personal communication. 1999. This hierarchical system was described at the WateReuse Symposium in Long Beach, CA as a triad of political, economic and technical factors. Mr. Kasower is Desalination Planning Manager with the US Bureau of Reclamation.

Keeney, R. and Raiffa, H. (1976) Decisions with Multiple Objectives: Preferences and Value Tradeoffs. (New York: Wiley)

Pinellas County Board of County Commissioners "Principles and Position Statements on Water Resources" (November 19, 2002) A full text is available at http://www.pinellascounty.org.

Pleasanton (CA) City Council. Minutes 12/7/99 http://www.ci.pleasanton.ca.us/archive/ccminutes991207.htm

Rosenblum, E. "Selection and Implementation of Nonpotable Water Recycling in Silicon Valley California, " *Water Science and Technology* Vol 40 No 4-5 pp 51–58, IWA Publishing (1999)

Rosenblum, E., T. Richardson and L. Byers. "Quantitative Evaluation of Water Reuse Alternatives in Silicon Valley." in *Decision Support Systems for Water Resources Management*, AWRA/UCOWR Conference (Snowbird, UT) pp. 233-238 (2000)

Rosenblum, E. and T. Richardson. "Benefit Cost Analysis Validates South Bay Water Recycling Value," in Proceedings, WateReuse California Section Annual Conference, February 26-28, 2003 (San Francisco, CA).

Sabatier, Paul. (1988) "An advocacy coalition framework of policy change and the role of policy-oriented learning therein." Policy Sciences 21:129--168 (Kluwer Academic Publishers, Dordrecht, The Netherlands)

Sheikh, B., E. Rosenblum et al. (1998) "Accounting for the Benefits of Water Reuse," In *Proceedings*, AWWA/WEF 1998 Water Reuse Conference; Orlando, FL

VanWagoner, W., T. Snow, B. Butler, B. Bailey, and T. Richardson. "Use of Recycled Water to Augment Potable Supplies: An Economic Perspective." A White Paper by the Potable Reuse Committee, WateReuse Association (T. Richardson, Chair).

Wegner-Gwidt, Joyce. "Public Support and Education for Water Reuse," in *Wastewater Reclamation and Reuse*, Asano, T. ed. (Technomic Publishing Co., Lancaster, PA) 1998. pp.1432-1438

Webb, Erik K. Personal communication. 2004. The significance of relative criticality as an index of consensus was described at the WateReuse Symposium in Phoenix, Arizona (September 19-22, 2004). Dr. Webb is a Congressional Fellow with the United States Senate Committee on Energy and Natural Resources.

Keyword Index

IWA Publishing's authorised EU representative for General Product Safety Regulations is Diane D'Arras, 15 rue Duret, 75116 Paris, France, e-mail: safety@iwap.co.uk.

Printed and bound by CPI Group (UK) Ltd, Croydon, CR0 4YY
16/06/2026
02140344-0002